Princeton Review

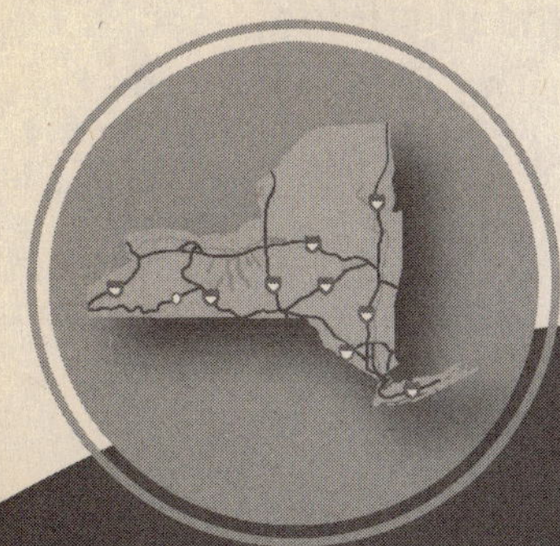

Roadmap to 4th Grade Math

NEW YORK EDITION

by
Diane Perullo

Random House, Inc.
New York

www.princetonreview.com

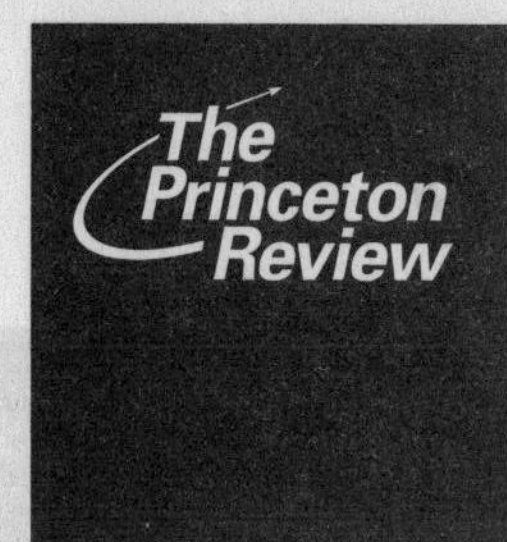

This workbook was written by The Princeton Review, one of the nation's leaders in test preparation. The Princeton Review helps millions of students every year prepare for standardized assessments of all kinds. The Princeton Review offers the best way to help students excel on standardized tests.

Princeton Review Publishing, L.L.C.
160 Varick Street, 12th Floor
New York, NY 10013

E-mail: textbook@review.com

Published in the United States by Random House, Inc., New York.

ISBN 0-375-76353-8

Editor: Linda Fan
Director of Production: Iam Williams
Design Director: Tina McMaster
Development Editor: Scott Bridi
Art Director: Neil McMahon
Production Manager: Mike Rockwitz
Production Coordinator: Robert Kurilla
Production Editor: Wade Ostrowski
Ollie the Ostrich illustrated by Paulo De Freitas, Jr.

Manufactured in the United States of America

9 8 7 6 5 4 3 2 1

First Edition

Acknowledgements

This book is the result of a team effort with my editor, Linda Fan. Her insights, suggestions, and creativity were a tremendous help to me. Many thanks, Linda. Thanks also to Russell Kahn for his continuing efforts on my behalf, and to Scott Bridi for quick responses and clear guidance. Finally, thanks to the production team of Wade Ostrowski and Robert Kurilla.

This book is dedicated with love to Jim, Bill, and Donna.

Contents

INTRODUCTION

Introduction for Parents and Teachers

About The Princeton Review

The Princeton Review is one of the nation's leaders in test preparation. We prepare more than two million students every year with our courses, books, on-line services, and software programs. In addition to helping New York State students with their grade 4 and grade 8 tests, we make study guides for the New York high school Regents series. We also coach students around the country on their statewide standardized tests and on college entrance exams such as the SAT-I, SAT-II, PSAT, and ACT. Our strategies and techniques are unique and, most importantly, successful. Our goal is to reinforce skills that students have been taught in the classroom and show them how to apply those skills to the specific format and structure of the tests in the New York State Testing Program.

About This Book

Roadmap to 4th Grade Math: New York Edition contains three main parts: an introduction, lessons in which students practice individual math skills, and practice tests. The introduction (both this parent/teacher introduction and the student introduction) explains the particulars of the New York State Grade 4 Mathematics test, including vital information such as the tools students can use while taking the test and the amount of time they will have to finish the test.

Each lesson (or "Mile") focuses on a specific math skill that is included in the New York State fourth-grade math curriculum. The thirty-three Miles in this book include the skills that will be tested on the New York State Grade 4 Math test. The first two Miles provide techniques that students may use to answer the different types of questions that will appear on the test: multiple choice, short response, and extended response. An answer key for activities and items in the lessons begins on page 81.

You can use the Progress Chart on pages 10 and 11 to encourage and motivate students. The chart shows the sequence of the thirty-three Miles included in this book. Have students color in or cross out each mile marker as they complete the Miles. As students work through the book, remind them that they are doing a good job and making progress toward reaching their goals.

This book also contains two full-length practice tests. Each practice test is modeled after the New York State Grade 4 Math test in terms of length, format, question types, and skills assessed. Reviewing students' performance on the tests will help you assess which skills they need more practice with before they take the New York State Grade 4 Math test. Instructions for administering the practice tests appear on page 100. Answer keys and explanations of correct answers for the two practice tests begin on page 181.

Before students begin working through this book, look through the content of the Miles. Every student, and every class, has different strengths and weaknesses. It's perfectly fine if students work on the lessons out of sequence.

About the New York State Testing Program

All public school students in New York State take tests in mathematics, English language arts (ELA), and science in grades 4 and 8. Students in grades 5 and 8 also take a test in social science. Test results are used in several ways. Teachers and school administrators use the test scores as a factor in determining whether a student should be promoted to the next grade. Test scores can also be a factor in deciding if a student needs extra instruction during school, tutoring after school, or summer school. In fact, students who score below Level 3 on the Grade 4 Math or ELA tests will receive "instructional intervention," according to the New York State Education Department. The type of intervention varies from one school district to another.

About the New York State Grade 4 Mathematics Test

The New York State Grade 4 Mathematics test is administered over a three-day period in early May. One session of the test is administered on each of the three days. Check your school's schedule for the exact testing dates.

Session 1 of the Grade 4 Math test includes about thirty multiple-choice questions. All multiple-choice questions must be answered on a separate answer sheet. Students will have forty minutes to answer the questions in Session 1.

Session 2 and Session 3 each include eight to ten open-response questions. Open-response questions can be two-point short-response questions or three-point extended-response questions. In each of the two sessions, students will have fifty minutes to answer the questions. Halfway through each session, they will have a short break that does not count against the time allotted for the sessions. Responses to the questions in Session 2 and Session 3 should be written directly in the test book.

During all three sessions of the test, students are allowed to write in the test book. This may be an advantage for the students because it allows them to write their calculations right next to the problems they are solving.

Students will be given punch-out tools to use while they take the test. The tools include a ruler that measures inches and centimeters, a set of counters, and a set of pattern blocks. We've provided similar tools for students on page 7. Cut out these tools with scissors so that students can use them as they work on the Miles and practice tests in this book.

Students with disabilities may be allowed certain special testing accommodations. Information concerning test access and modification for students with disabilities may be found on the Internet at ftp://unix2.nysed.gov/pub/education.dept.pubs/vesid/oses/test.access.mod/testacce.txt.

For additional information about testing and curricula, visit the New York State Education Department's Web site: www.emsc.nysed.gov.

Introduction for Students

About This Book

This book is designed to help you improve your math grades. It will also help you improve your score on the New York State Grade 4 Math test, which you will take in the spring. In this book, you will review and practice the important skills that you should know for your fourth-grade math class. These are the same skills that you need to know for your math test. You can learn to do better on the test and in your class at the same time.

There are three main parts of this book: this introduction, the Miles, and the practice tests. In the Miles, you will review and practice individual math skills. The practice tests give you a chance to take tests that are similar to the actual New York State Grade 4 Math test.

About the New York State Grade 4 Math Test

All fourth-grade students in New York State public schools have to take the Grade 4 Math test. The purpose of the test is to show your teachers and parents what you know. The results of the test will tell them which math skills you know well and which ones you might need a little help on.

You will take the test during a three-day period sometime in May. (You can ask your teacher for the exact testing dates if you don't know them.) On the first day of testing, you will answer about thirty multiple-choice questions in forty minutes. On each of the next two days of testing, you will answer about nine open-response questions in fifty minutes.

About the Multiple-Choice Questions

Multiple-choice questions have four answer choices (A, B, C, and D or F, G, H, and J). To answer these questions, you need to pick the best answer choice and mark it with your pencil on the separate bubble sheet. Make sure to answer every question, because you can't earn points for a blank answer. Check out Mile 1 (on pages 14 and 15) for more information about answering this type of question.

About the Open-Response Questions

Open-response questions require you to write out your final answer and all the work you did to get the answer. For the open-response questions, you can earn partial credit even if your final answer is incorrect. That's why it's very important to show your work. Write clearly. If the grader can't read your handwriting, you may lose out on points. Check out Mile 2 (on pages 16 and 17) for more information about answering this type of question.

Tools

You will be given a ruler, counters, and pattern blocks to help you answer the questions on the test. On page 7, there are copies of these tools that are similar to the ones you will use on the test.

As you work on the Miles and practice tests in this book, you will need the tools on page 7. There are icons and instructions throughout the book that tell you when you will need to use these tools. You may want to tape an envelope to the cover of this book and store the tools in there.

Preparing for the New York State Grade 4 Math Test

Here is a list of things you can do to prepare for the New York State Grade 4 Math test.

- **Ask questions.** If you are confused after you finish working on a Mile (or even while answering just one question), ask a parent or teacher for help. Asking questions is the best way to make sure that you understand what you have to do in order to do well on the test.

- **Practice.** Math is all around you. Every time you go to the grocery store, make dinner, or play sports, math comes along for the ride. Use real-life opportunities to practice the math you know. Estimate the prices of the groceries in your cart while you're waiting on line. See if your estimation comes close to the register sum. If you play softball or baseball on a team, keep track of your batting average. Every day there will be more math to do, and the great thing is that it's the same type of math that will be on the Grade 4 Math test you'll be taking.

- **Read.** Read everything you can. Read the newspaper, magazines, books, plays, poems, comics, and even the back of your cereal box. The more you read, the better you will become at it. And the better you read, the more likely you will be to do well on the Grade 4 Math test. Many of the questions on the math test will be word problems. If you can't read and understand the problems, you won't have a chance to show all the math you know.

- **Eat well and get a good night's sleep.** Your body doesn't work well when you don't eat good food and get enough sleep. Neither does your brain. On the night before the test, make sure to go to bed at your normal time and get plenty of sleep. You should also eat a healthy breakfast on the morning of the testing days. It is important to be awake and alert while you take the Grade 4 Math test, or any other test for that matter.

This is just the beginning of the road. There are great things to learn ahead. So buckle your seat belt and get ready to travel the first mile to math excellence.

Cut-Out Tools

Ruler

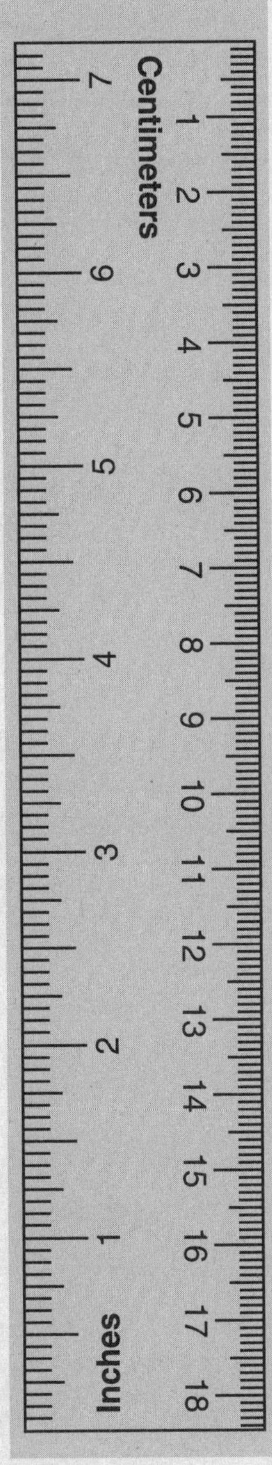

Counters

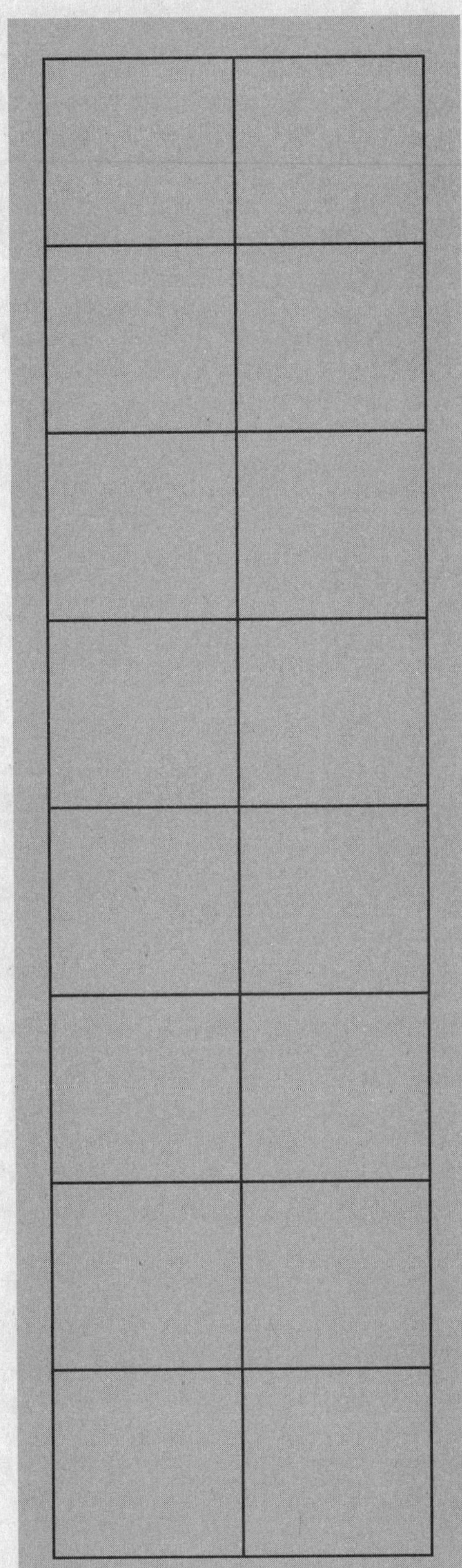

Pattern Blocks

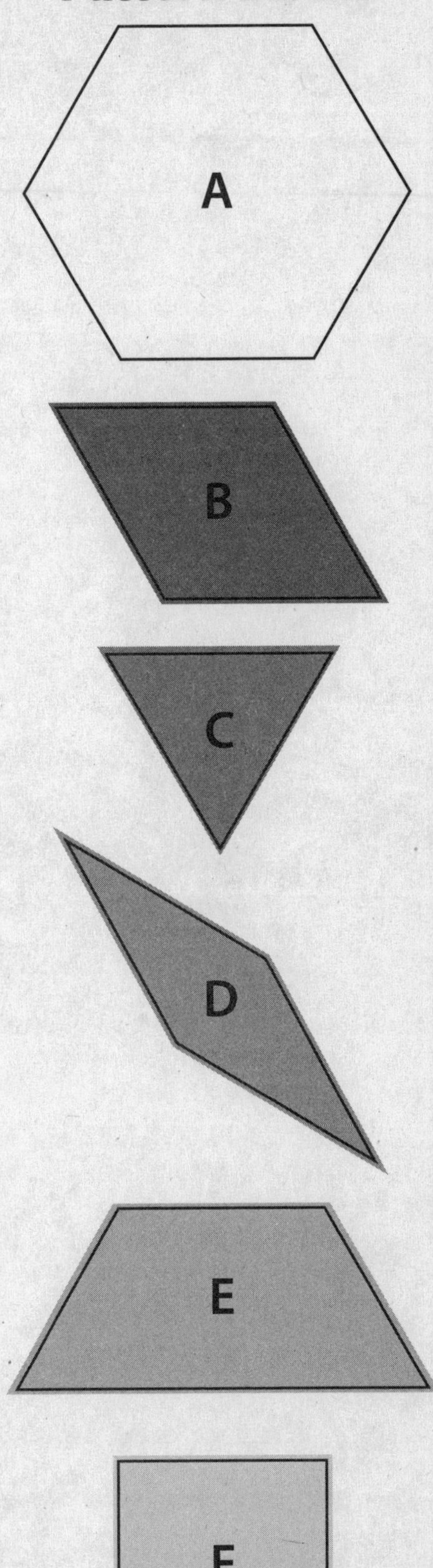

Progress Chart

Progress Chart

Directions: Cross off or color in each mile marker after you complete each activity.

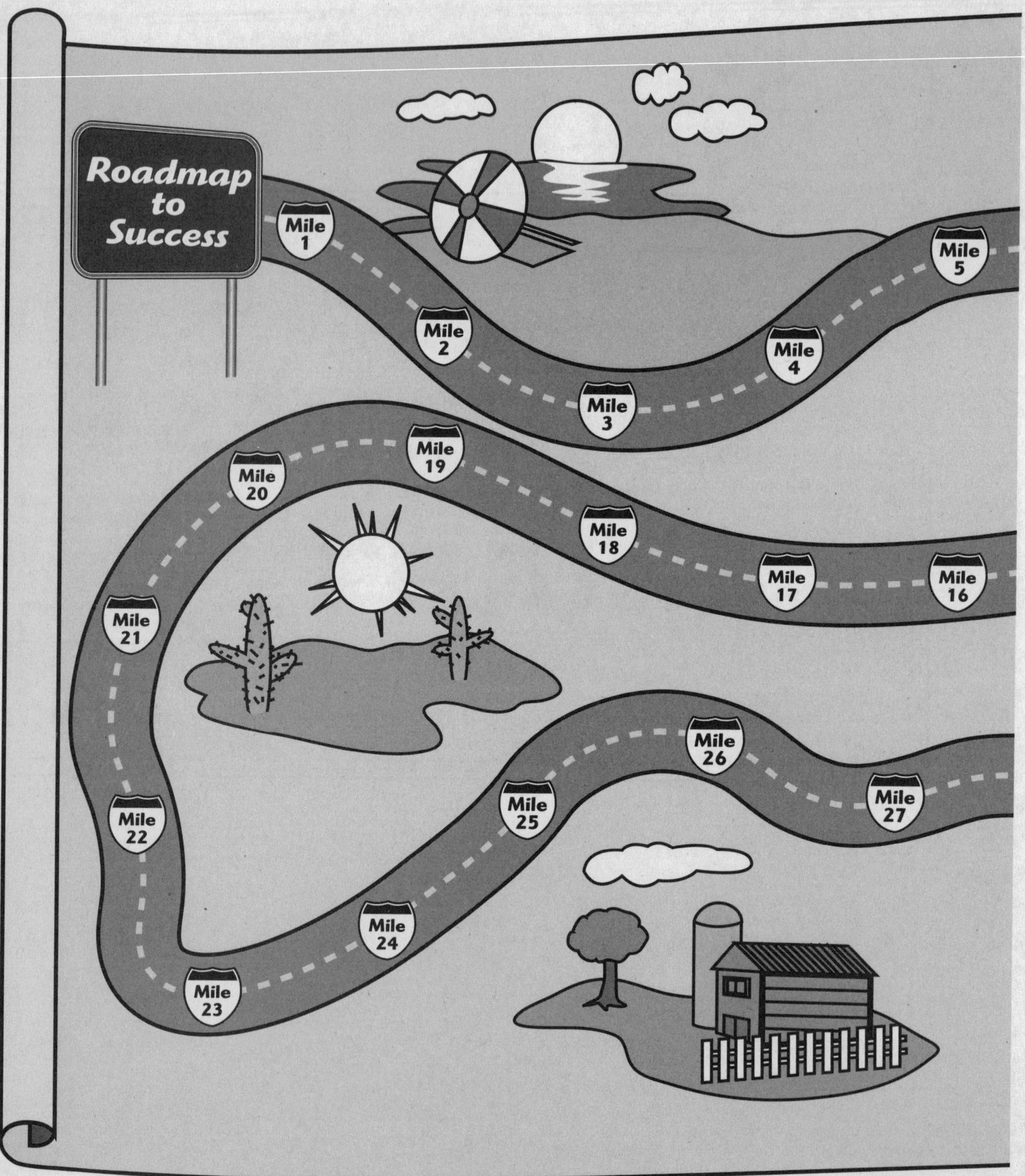

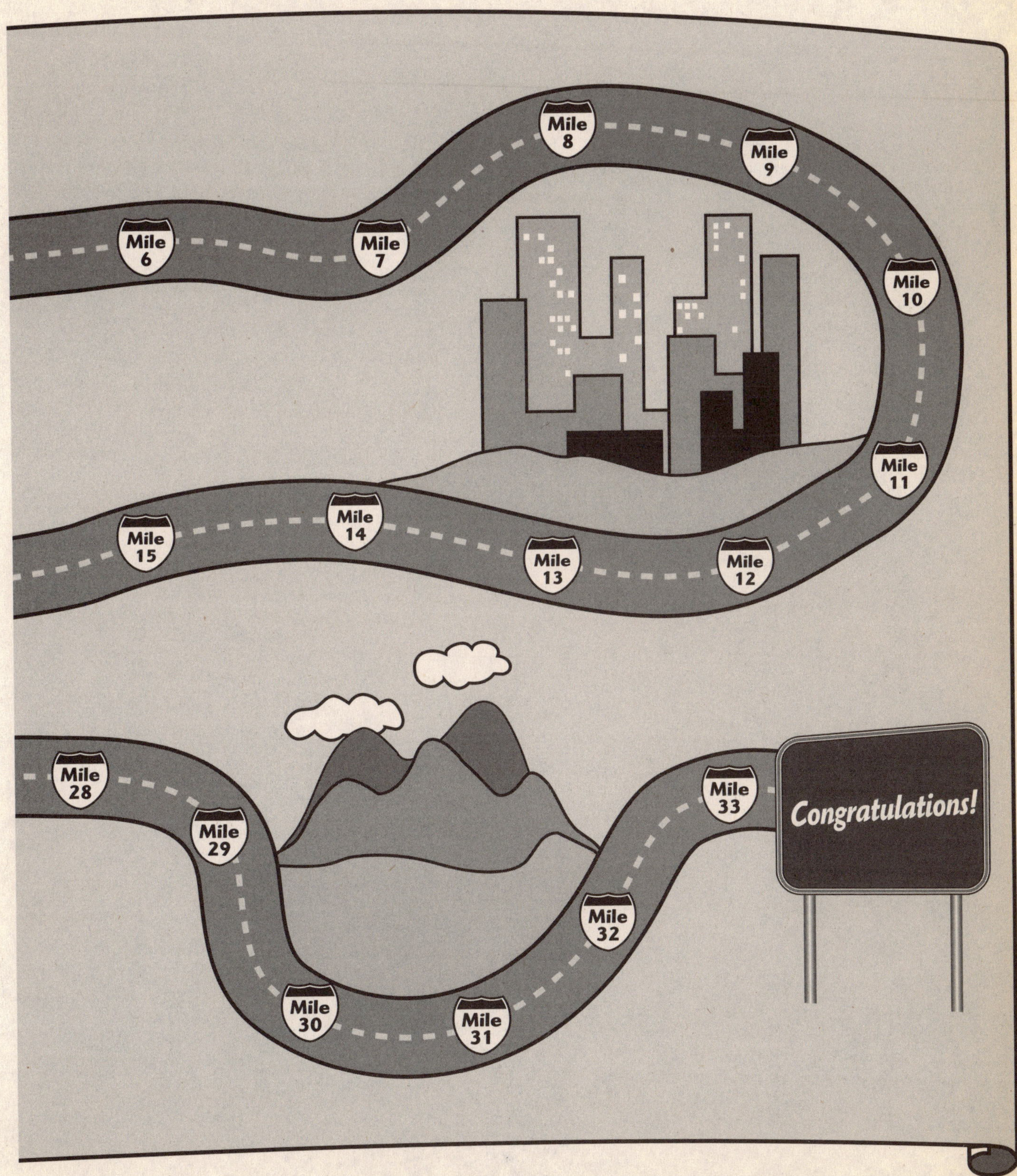
Mile 6
Mile 7
Mile 8
Mile 9
Mile 10
Mile 11
Mile 12
Mile 13
Mile 14
Mile 15
Mile 28
Mile 29
Mile 30
Mile 31
Mile 32
Mile 33
Congratulations!

Mile by Mile

MILE 1: ANSWERING MULTIPLE-CHOICE QUESTIONS

The New York State Grade 4 Math test consists of about forty-eight questions, thirty of which are multiple-choice. Each multiple-choice question is followed by four possible answer choices. The choices could be A, B, C, and D or F, G, H, and J. You have to pick the best answer choice for each question.

You should approach multiple-choice questions like any other math problems. First, carefully read the question and figure out what you need to do in order to answer it. You may have to perform calculations or use reasoning to figure out the answer. What makes multiple-choice questions different is that they come with four answer choices, one of which is the correct answer. Once you figure out your answer, look for it among the answer choices. Once you find it, just fill in the bubble that goes with it.

But what happens if it's not as easy as we made it sound? What if you have trouble figuring out the answer to a question? That's where **Getting Rid of Wrong Answers** comes in. First, look for the choices that you *know* are wrong. Then, cross them out. Get rid of them!

Try this math problem.

▶ **17 + 68 =**

17
+68
85

A 51

B 65

C 75

D 85

Solve to find the answer. 17 + 68 = 85. Right!

Did you know the answer? If not, it doesn't mean that you can't pick the correct answer. Just get rid of some of the answers that you know are wrong. Then you can choose from the ones that are left.

Look at the answer choices again. You know the question is an addition problem, so the sum must be greater than 17 and 68. Answer choice **A** can't possibly be the correct answer, because it is less than 68. Get rid of it! Cross it off with your pencil. Answer choice **B** is less than 68. Get rid of it too! Answer choice **C** seems possible, so hold on to it. Answer choice **D** is possible too. Now you know that the correct answer must be answer choice **C** or **D**. If you are still stuck, take a guess.

What if you can't get rid of all of the wrong answer choices? That's all right. If you can get rid of even one or two choices, you have a better chance of picking the right answer. This can help you on any test you take that has multiple-choice questions.

Look at the following example to see how you can get rid of wrong answer choices on the New York Grade 4 Math test.

► **Ella bought 5 packages of daffodil bulbs. Each package had 30 bulbs. How many bulbs did Ella buy in all?**

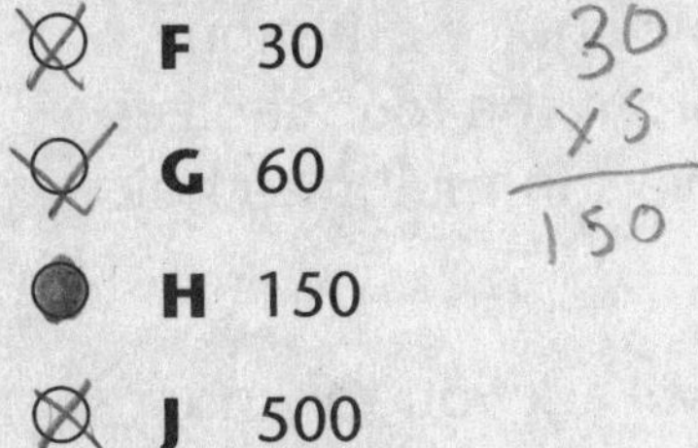

F 30

G 60

H 150

J 500

If you are having trouble solving this, try to get rid of the answer choices that you know are wrong.

Look at **F**. This answer is much too small. Ella bought several packages and each had 30 bulbs. It would mean that Ella bought only one package. This answer isn't right. Get rid of **F**.

Look at **G**. This answer still seems too small. Get rid of **G**.

Look at **H**. This answer seems reasonable. Don't do the math yet. Keep **H** and come back to it.

Look at **J**. Ella did not buy 500 bulbs. Ella only bought 5 packages of 30 bulbs each, which doesn't give her nearly 500. Get rid of choice **D**.

That leaves you with choice **H**. Check to make sure the answer is correct by multiplying the 30 bulbs by 5 packages. $30 \times 5 = 150$. You found the correct answer by eliminating the answer choices that do not make sense!

You probably won't find many correct answers without doing the math. But getting rid of wrong answer choices can be very helpful when you get stuck on a question. Even if you can get rid of only one or two choices, your test scores can go up.

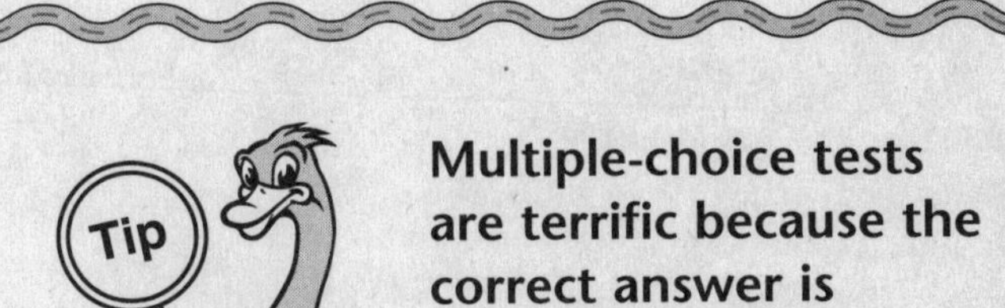

Multiple-choice tests are terrific because the correct answer is printed on the page. You just have to choose the right answer choice!

Mile 2: Answering Short-Response and Extended-Response Questions

Session 1 of the New York State Grade 4 Math test contains multiple-choice questions. In that session, you will be able to pick your answer to a question from the four choices you are given. In Mile 1, you learned how to get rid of wrong answers in multiple-choice questions.

Sessions 2 and 3 of the Grade 4 Math test contain questions that will ask you to write your own answers. These are the **short-response** and **extended-response questions.**

Don't let the names *short-response* and *extended-response* confuse you. The questions look the same. Only the point value of the questions is different. A short-response question is worth 2 points. An extended-response question is worth 3 points. You won't know which is a 2-point question and which is a 3-point question on the test, so don't worry about the difference.

Show **all** of your work for short-response and extended-response questions. It is important that you show how you got the answer. If you do not show your work, you will not receive full credit for that question, even if you gave the correct answer.

Often, the test will have the instruction ***"Show your work"*** on the page, with space provided. That's a helpful reminder, but even if the test doesn't ask, you should show your work somewhere on the page.

Try a couple of questions so you can see how to answer short-response and extended-response questions.

► **There is going to be a family party at Jamie's house. There will be 17 adults and 25 children attending the party. Jamie wants to make 4 of her special oatmeal cookies for each person. How many cookies does Jamie need to make in all?**

Show your work.

17 + 25 = 42 people   42 × 4 = 168 cookies

Answer 168 **cookies**

The first thing you would need to do is to find out how many people will attend the party. Under ***"Show your work,"*** add the number of adults and children.

$$\begin{array}{r} 17 \text{ adults} \\ +\ 25 \text{ children} \\ \hline 42 \text{ people} \end{array}$$

Label your calculations so you know what they mean.

Next, multiply the number of people by 4 to get the total number of cookies Jamie would need to make. Under ***"Show your work,"*** you would also write

$$\begin{array}{r} 42 \text{ people} \\ \times\ 4 \text{ cookies} \\ \hline 168 \text{ cookies} \end{array}$$

Finally, on the line next to ***"Answer,"*** write 168.

If you showed all of your work and wrote the correct answer, you would receive full credit for the question. You would receive partial credit if you got the correct answer but did not show your work, or if you got the incorrect answer but showed some correct calculations.

Try another one.

▶ **Your teacher wrote the following number sentence on the board:**

12 + 9 + 25 = 25 + 9 + 12

Is this number sentence correct? Explain how you know if this sentence is correct or not without solving it.

This number sentence is correct because it is commutative property. A communitive property is when numbers orders are changed up but still they have all the same numbers

First, think about the number sentence. Adding the parts of an addition problem (addends) in different orders does not change the sum. That means the number sentence is correct. This is an example of the commutative property.

Next, write that the sentence is correct and explain your reason on the lines provided. A good answer might be: The number sentence is correct because of the commutative property. Changing the order of the addends does not change the sum.

If you answered that the number sentence is correct and gave your reason, you would receive full credit for this question. Good job!

Mile 3: Place Value

You know that 300 is larger than 30. Why? Both numbers have the digit 3, but the 3 is in a different place value for each number. In the number 300, the digit 3 is in the hundreds place, which means that the 3 stands for three hundreds. In the number 30, the digit 3 is in the tens place, which means that the 3 stands for three tens. Hundreds are greater than tens, so 300 is bigger than 30.

A number with six digits, like 613,845, has six place values. You name the place values from the ones place. From right to left, the **place values** are ones, tens, hundreds, thousands, ten thousands, and hundred thousands.

Directions: Match the numbers with their correct written forms. The letters will spell the last name of the fourth U.S. president, which is also the name of the capital of Wisconsin. Here's a hint: Think about the place values of the numbers to help you complete this activity.

1. M	thirty-seven thousand, four hundred thirty	S)	6,402
2. A	five hundred twenty-five thousand, one hundred seventy-two	D)	178
3. D	one hundred seventy-eight	N)	812,651
4. I	four thousand, nine hundred twenty-eight	M)	37,430
5. S	six thousand, four hundred two	O)	78,635
6. O	seventy-eight thousand, six hundred thirty-five	I)	4,928
7. N	eight hundred twelve thousand, six hundred fifty-one	A)	525,172

► **The last name of the fourth president and the capital of Wisconsin is** Madison.

Directions: Read each question carefully. Fill in the bubble next to your answer choice for each question.

1 In 2000, the population of Albany was 95,658. Which statement is true about the number 95,658?

 **A** The digit 9 is in the thousands place.

 **B** There are 5 ten thousands.

● **C** The digit in the ten thousands place is odd.

 **D** The value of the 6 is 6,000.

2 The length of Long Island is about 623,040 feet. What digit is in the ten thousands place?

 **F** 6

 **G** 2

 **H** 3

 **J** 4

3 What is 49,576 in written form?

 **A** Forty-nine thousand, five hundred seventy

 **B** Four hundred ninety thousand, five hundred seventy-six

 **C** Four thousand, ninety-five hundred seventy-six

● **D** Forty-nine thousand, five hundred seventy-six

4 The area of Mexico is 756,066 square miles. Which statement is false about the number 756,066?

● **F** A 6 is in the ten thousands place.

 **G** The 0 is in the hundreds place.

 **H** There are 7 hundred thousands.

 **J** A 6 is in the tens place.

Directions: Read each question carefully and write your answer on the line provided.

5 In the number 875,032, what digit is in the ten thousands place?

Answer 7

6 Write the number three hundred twenty-nine thousand, five in standard form.

Answer 329,005

Zeros can be very important. By itself, a zero may be worth nothing, but in a larger number, it holds a place value.

MILE 4: COMPARING AND ORDERING NUMBERS

When **comparing** and **ordering numbers,** you see how numbers relate or compare to each other. You can see which numbers are greater than or less than other numbers. You can compare and order whole numbers or decimals.

Look at the numbers 567 and 576. To find out which number is greater, stack the numbers, lining up the place values. Then, compare the digits in each place from left to right.

567
576

The first digit that is different is in the tens place. Compare the 6 and the 7. Because 7 is greater than 6, 576 is greater than 567, or $576 > 567$. You do not need to compare the remaining digit.

Use the correct symbols when comparing numbers. Remember that the small end of the symbol points to the smaller number.

The symbol $>$ means greater than.

The symbol $<$ means less than.

Directions: The letters below are equal to a specific decimal place on the number line. Put the letters on their correct place in line to spell the name of a wild dog found in Australia.

N = 7.1

I = 6.7

G = 7.6

D = 6.3

O = 7.9

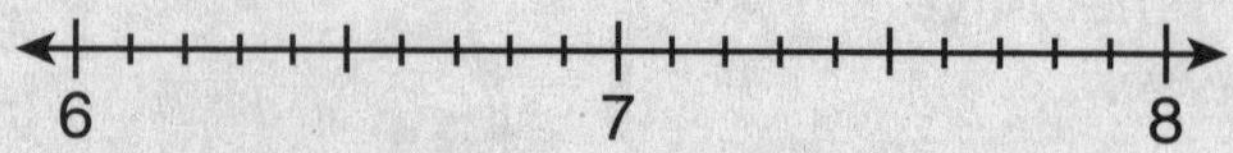

▶ **I am a small wild dog found in Australia. What am I?**

____

Directions: Read each question carefully. Fill in the bubble next to your answer choice for each question.

1 Letty watched the street numbers increase as she rode her bus to school. She passed 149th Street. What would be the next street?

- ○ **A** 151st ST
- ○ **B** 159th ST
- ○ **C** 150th ST
- ○ **D** 249th ST

2 Which arrow on the number line below is pointing to 14.7?

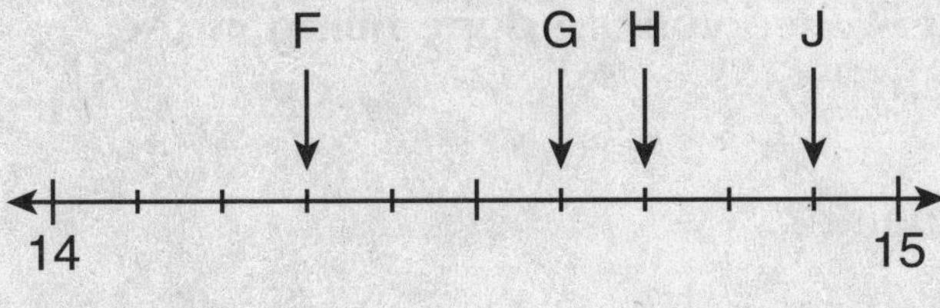

- ○ **F**
- ○ **G**
- ○ **H**
- ○ **J**

3 Which number should be written in the box to make the number sentence true?

$28 + 9 > \square + 22$

- ○ **A** 14
- ○ **B** 15
- ○ **C** 16
- ○ **D** 17

4 Which of the following statements is true?

- ○ **F** $25.5 > 25.6$
- ○ **G** $23.6 > 23.0$
- ○ **H** $2.45 = 24.5$
- ○ **J** $25.0 < 25$

Directions: Fill in the boxes with numbers that will make the expressions true.

5 $894 < 896 < \square < 900$

6 $10.5 > 10.4 > \square > 10.2$

MILE 5: ADDITION

Addition is the combining of the values of two or more numbers. These numbers are called **addends.** After you combine the numbers, the new number is called the **sum.**

Directions: Add the following pairs of numbers. Each sum goes with the letter next to it. You will use the sums to answer the question at the bottom of the page.

1. $\begin{array}{r} 284 \\ +\ 257 \\ \hline \end{array}$ (G)

2. $\begin{array}{r} 426 \\ +\ 702 \\ \hline \end{array}$ (S)

3. $\begin{array}{r} 194 \\ +\ 23 \\ \hline \end{array}$ (A)

4. $\begin{array}{r} 860 \\ +\ 184 \\ \hline \end{array}$ (L)

5. $\begin{array}{r} 536 \\ +\ 368 \\ \hline \end{array}$ (U)

6. $\begin{array}{r} 339 \\ +\ 151 \\ \hline \end{array}$ (X)

7. $\begin{array}{r} 246 \\ +\ 272 \\ \hline \end{array}$ (T)

8. $\begin{array}{r} 608 \\ +\ 373 \\ \hline \end{array}$ (Y)

Answer the question below using the letters next to each addition problem. Write the letter that goes with each sum in the boxes below. You will not use all of the letters, and you can use letters more than once. When you are done filling in the boxes, you will have the answer to this question:

► **I am a large group of stars. What is my name?**

___	___	___	___	___	___
541	217	1,044	217	490	981

Remember to check your answers in the answer key, which starts on page 81.

Directions: Read each question carefully. Fill in the bubble next to your answer choice for each question.

1 485
\+ 321

○ **A** 706

○ **B** 796

○ **C** 806

○ **D** 816

2 1,729
\+ 534

○ **F** 1,253

○ **G** 1,273

○ **H** 2,253

○ **J** 2,263

3 23
\+ 98

○ **A** 101

○ **B** 110

○ **C** 121

○ **D** 131

4 564
\+ 307

○ **F** 861

○ **G** 871

○ **H** 961

○ **J** 971

5 1,106
\+ 844

○ **A** 1,840

○ **B** 1,950

○ **C** 2,040

○ **D** 2,050

6 267
\+ 683

○ **F** 950

○ **G** 940

○ **H** 850

○ **J** 840

Always remember to carry the 1 (or regroup) when two digits in a column add up to 10 or greater.

Mile 6: Subtraction

Subtraction is a friendly type of math because it allows you to **borrow.** Just like you borrow a sweater or a CD from a friend, you can borrow from different columns of numbers when you subtract. *You have to borrow when the top number in a column is smaller than the number below it.* The answer to a subtraction problem is called the **difference.**

Directions: Calculate the following problems. Draw a line from the baseball to the mitt with the correct answer.

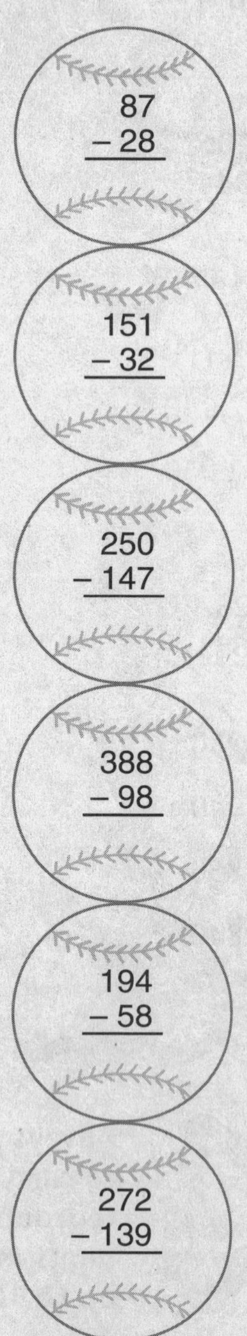

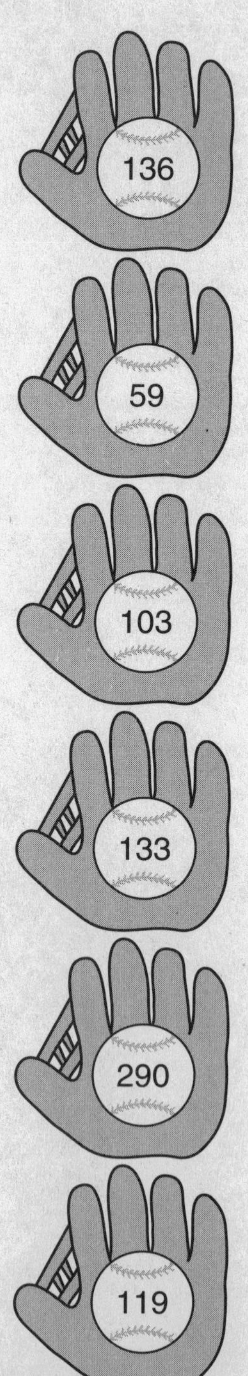

Directions: Read each question carefully. Fill in the bubble next to your answer choice for each question.

1 458
− 82

○ **A** 266
○ **B** 366
○ **C** 376
○ **D** 476

2 67
− 29

○ **F** 36
○ **G** 38
○ **H** 42
○ **J** 46

3 607
− 234

○ **A** 263
○ **B** 353
○ **C** 363
○ **D** 373

4 259
− 83

○ **F** 7
○ **G** 166
○ **H** 176
○ **J** 186

5 145
− 139

○ **A** 6
○ **B** 9
○ **C** 15
○ **D** 16

6 750
− 226

○ **F** 474
○ **G** 514
○ **H** 524
○ **J** 534

You can check your answer after you subtract. Add your answer to the subtracted number in the problem. The two numbers should add up to the original number.

Mile 7: Multiplication

Multiplying numbers is the same as repeating addition. For example, you can think of the expression 5×3 as adding five 3s together, $3 + 3 + 3 + 3 + 3$, or adding three 5s together, $5 + 5 + 5$. The answer will be the same either way, 15. The numbers you multiply, 3 and 5, are called **factors**. The answer, 15, is called the **product**.

For more complex problems, you should stack the two numbers, placing the larger number on top. Then, multiply each digit of the top number by the bottom number separately. Carry if necessary.

$$\begin{array}{r} 41 \\ \times\ 6 \\ \hline 246 \end{array}$$

Directions: Multiply these numbers. Then use the code box at the bottom of the page to find a letter for each product. The letters should be placed in the same order as the multiplication problems below. When you put all of the letters in order, you will spell out the name of a famous mathematician.

1. $\begin{array}{r} 18 \\ \times\ 9 \\ \hline \end{array}$ **2.** $\begin{array}{r} 92 \\ \times\ 5 \\ \hline \end{array}$ **3.** $\begin{array}{r} 37 \\ \times\ 3 \\ \hline \end{array}$ **4.** $\begin{array}{r} 75 \\ \times\ 7 \\ \hline \end{array}$ **5.** $\begin{array}{r} 77 \\ \times\ 4 \\ \hline \end{array}$

6. $\begin{array}{r} 69 \\ \times\ 6 \\ \hline \end{array}$ **7.** $\begin{array}{r} 56 \\ \times\ 2 \\ \hline \end{array}$ **8.** $\begin{array}{r} 48 \\ \times\ 8 \\ \hline \end{array}$ **9.** $\begin{array}{r} 83 \\ \times\ 5 \\ \hline \end{array}$ **10.** $\begin{array}{r} 24 \\ \times\ 9 \\ \hline \end{array}$ **11.** $\begin{array}{r} 16 \\ \times\ 4 \\ \hline \end{array}$

N 414	T 415	I 162	A 111	O 216	N 64
E 112	A 525	S 460	C 308	W 384	

▶ **What is the name of the famous mathematician?**

____

Directions: Read each question carefully. Fill in the bubble next to your answer choice for each question.

1
$$\begin{array}{r} 54 \\ \times\ 8 \\ \hline \end{array}$$

- ○ **A** 404
- ○ **B** 414
- ○ **C** 422
- ○ **D** 432

2
$$\begin{array}{r} 39 \\ \times\ 4 \\ \hline \end{array}$$

- ○ **F** 146
- ○ **G** 156
- ○ **H** 158
- ○ **J** 168

3
$$\begin{array}{r} 15 \\ \times\ 6 \\ \hline \end{array}$$

- ○ **A** 70
- ○ **B** 80
- ○ **C** 90
- ○ **D** 100

4
$$\begin{array}{r} 93 \\ \times\ 5 \\ \hline \end{array}$$

- ○ **F** 465
- ○ **G** 468
- ○ **H** 475
- ○ **J** 478

5
$$\begin{array}{r} 88 \\ \times\ 3 \\ \hline \end{array}$$

- ○ **A** 244
- ○ **B** 264
- ○ **C** 274
- ○ **D** 284

6
$$\begin{array}{r} 46 \\ \times\ 7 \\ \hline \end{array}$$

- ○ **F** 283
- ○ **G** 312
- ○ **H** 322
- ○ **J** 333

Division is the reverse of multiplication. You can use division to check your answers to multiplication problems. Divide the product by one of the factors. If the answer is the other factor, your multiplication is correct.

MILE 8: DIVISION

Division is an operation that separates something into equal parts. For example, if you have a classroom of 25 students that you want to break into groups of 4, how many groups will you have?

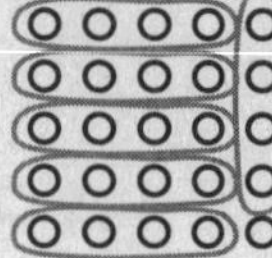

$$\begin{array}{r} 6\text{ r1} \\ 4\overline{)25} \\ \underline{-24} \\ 1 \end{array}$$

From the diagram, you can see that there will be 6 groups of 4 students and 1 student left over. The number that is being divided, 25, is called the **dividend,** and the number that is breaking the dividend into groups, 4, is called the **divisor.** The whole number answer, 6, is called the **quotient,** and any number left over after finding the quotient, 1, is the **remainder.**

Directions: Find the quotients and remainders for each of the division problems below. Each answer represents a letter. Match each answer with a box at the bottom of the page to answer a riddle. Not all the letters will be used.

1. (J) $27 \div 3 =$

2. (I) $36 \div 12 =$

3. (U) $48 \div 7 =$

4. (E) $18 \div 9 =$

5. (N) $64 \div 8 =$

6. (B) $24 \div 2 =$

7. (A) $7\overline{)86}$

8. (O) $5\overline{)71}$

9. (H) $3\overline{)129}$

10. (W) $8\overline{)251}$

11. (R) $7\overline{)217}$

12. (S) $9\overline{)97}$

► **Why was the archaeologist upset?**

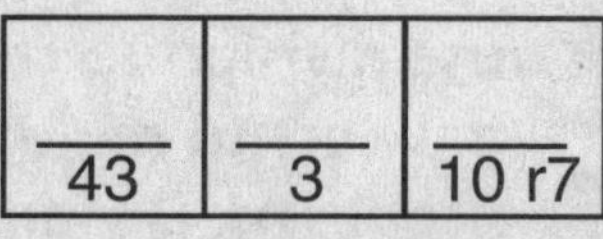

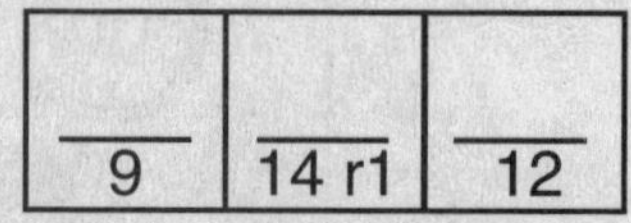

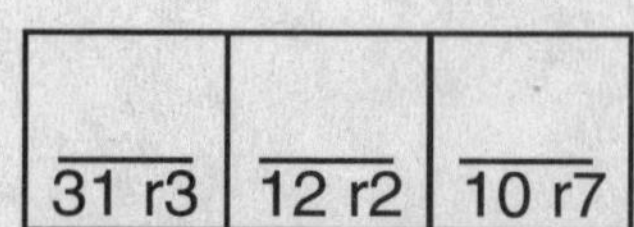

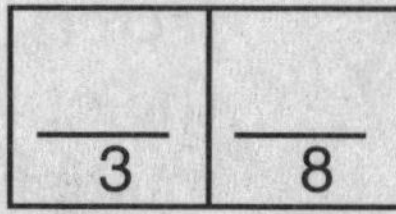

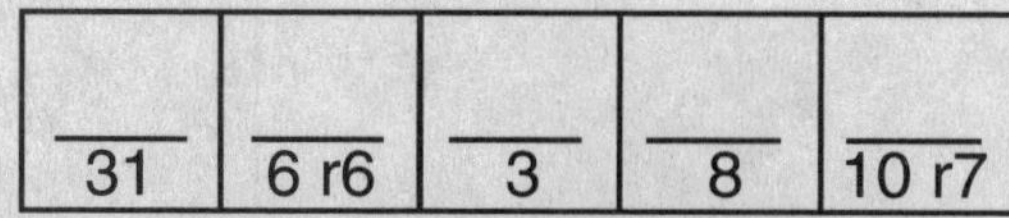

Directions: Read each question carefully. Fill in the bubble next to your answer choice for each question.

1 32 ÷ 8 = ?

○ **A** 8

○ **B** 6

○ **C** 4

○ **D** 2

2 Name the divisor in 45 ÷ 9 = 5.

○ **F** 5

○ **G** 9

○ **H** 15

○ **J** 45

3 72 ÷ 3 = ?

○ **A** 24

○ **B** 14

○ **C** 9

○ **D** 7

4 What is the remainder for 49 ÷ 6?

○ **F** 4

○ **G** 3

○ **H** 2

○ **J** 1

5 $4\overline{)144}$

○ **A** 28

○ **B** 30

○ **C** 34

○ **D** 36

6 25 ÷ 7 = ?

○ **F** 3 r3

○ **G** 3 r4

○ **H** 4 r1

○ **J** 4 r3

7 $7\overline{)150}$

○ **A** 20 r9

○ **B** 21 r2

○ **C** 21 r3

○ **D** 22 r1

The remainder is the amount that is left over after dividing. It will never be greater than the divisor.

MILE 9: MULTIPLES

Multiples are the products of a given whole number and any other whole number. In other words, the multiples of 2 are 2×1, 2×2, 2×3, 2×4, and so on. If you multiply, you'll see that some multiples of 2 are 2, 4, 6, 8, 10, 12, 14, and 16. You can also find multiples by counting by the number or skip-counting.

The multiples of 3 are 3, 6, 9, 12, 15, 18, 21, 24, and so on.

There is no limit to the number of multiples a number may have. As you can see, some numbers have multiples in common. The numbers 6 and 12 are multiples of both 2 and 3.

Directions: The numbers on the left are multiples of 4, 5, 6, or 7. Find the number that each is a multiple of. Then circle the letter beneath it. The letters from top to bottom will form the name of one of the seven continents.

	4	5	6	7
21	P	O	M	A
45	W	N	I	D
18	X	S	T	F
32	A	K	U	E
25	T	R	A	Y
49	Q	U	L	C
66	H	N	T	V
16	I	D	O	B
50	J	C	Z	I
54	S	E	A	R

► **What is the name of one of the seven continents?**

____

Directions: Read each question carefully. Fill in the bubble next to your answer choice for each question.

1 **Jack needs new pencils for school. There are 10 pencils in a complete package. If his mom bought a number of packages, how many pencils could she have bought?**

○ **A** 12

○ **B** 20

○ **C** 25

○ **D** 32

2 **Which of these numbers is a multiple of 3?**

○ **F** 31

○ **G** 32

○ **H** 35

○ **J** 36

3 **Leslie bought bottled water in packs of 6. If she wanted to buy only complete packs, how many bottles could she have bought?**

○ **A** 20

○ **B** 22

○ **C** 24

○ **D** 26

4 **Cecil delivers papers each morning. He sorts the papers into piles of 8. If he has no papers left over, how many papers did Cecil sort?**

○ **F** 49

○ **G** 52

○ **H** 54

○ **J** 56

5 **Patty is packaging strawberries that she has grown. She has 12 boxes. How many strawberries has she packaged, if she puts an equal number into each box?**

○ **A** 72

○ **B** 70

○ **C** 68

○ **D** 64

Mile 10: Operations

The four **operations** are addition, subtraction, multiplication, and division. There are different relationships between the operations.

- Addition and subtraction are opposite operations.
- Multiplication and division are opposite operations.
- Multiplication is repeated addition.
- Division is repeated subtraction.

Directions: Fill in each box with an operation sign (+, −, ×, ÷) to make each sentence true.

9 ☐ 5 = 45

18 ☐ 9 = 27

45 ☐ 9 = 5

27 ☐ 9 = 18

Directions: Look at the number sentences below. Write a sentence that uses the same three numbers but uses a different operation.

1. 5 × 8 = 40 ____________________

2. 54 − 13 = 41 ____________________

3. 23 + 11 = 34 ____________________

4. 7 × 12 = 84 ____________________

5. 19 − 7 = 12 ____________________

6. 36 ÷ 2 = 18 ____________________

Directions: Read each question carefully. Fill in the bubble next to your answer choice for each question.

1 Look at the number sentence below.

$36 \times 24 = 884$

Which expression can you use to check if the number sentence is correct?

○ **A** $884 - 24$

○ **B** $884 + 24$

○ **C** $884 \div 24$

○ **D** 884×24

2 Mrs. Johnson filled 6 party bags. In each bag, she placed 4 pieces of fruit. Which of the following shows how many pieces of fruit Mrs. Johnson used?

○ **F** $6 + 6 = 12$

○ **G** $6 + 4 = 10$

○ **H** $4 + 4 + 4 + 4 = 16$

○ **J** $4 + 4 + 4 + 4 + 4 + 4 = 24$

3 Which of the following is the same as

$24 \times 17 \times 6 \times 11$

○ **A** $11 + 6 + 17 + 24$

○ **B** $11 \times 6 \times 17 \times 24$

○ **C** $24 + 17 + 6 + 11$

○ **D** $24 \times 17 \times 17$

Directions: Fill in the boxes for each question.

4 Use the numbers 6 and 18 to fill in the boxes and make the number sentences correct.

$\square \times 3 = \square$

$\square \div 3 = \square$

5 Use the numbers 13 and 8 to fill in the boxes and make the number sentences correct.

$\square + 5 = \square$

$\square - 5 = \square$

Mile 11: Word Problems

Directions: Read the story and solve the problems as you read them.

The fourth-grade international club decided to host a lunch to raise money for club activities. Kathy, the club president, was chosen to organize the lunch. She asked if anyone wanted to volunteer to make or bring in something to sell.

Doug said that he would bring in 3 pizzas. Patty and Amy each agreed to make 1 dozen pork dumplings. Karl said he could bring in 2 dozen vegetable dumplings.

1. *How many individual dumplings will be brought in?* ____________

"Anybody else?" asked a very pleased Kathy. Everybody seemed so excited by the lunch.

Ms. Lunnette, the club advisor, spoke up. "I can make my famous tacos. We can also sell cups of tea and lemonade to go with lunch." Everyone thought that was a great idea.

Kathy decided to bring in fried rice that could be spooned into small bowls.

2. *If Kathy brings in 64 ounces of fried rice and sells it in 8-ounce bowls, how many servings will there be for sale?* ____________

On the day of the lunch sale, the students brought their goodies. Prices were set, with Ms. Lunnette's tacos being the most expensive at $2 each.

3. *If every taco was sold and it made $28 for the international club, how many tacos were there?* ____________

Doug was pleased to see that most of his 3 pizzas were sold. There was a slice of cheese and a slice of pepperoni left at the end of the day.

4. *If Doug sold 22 slices of pizza and each pizza was cut into the same number of pieces, how many pieces was each pizza divided into?* ____________

After the sale, Karl rubbed his stomach and moaned, "I think I ate all but 3 of Amy's dumplings."

5. *How many dumplings does Karl think he ate?* ____________

The sale was a great success! The international club had raised a lot of money for its activities.

Directions: Read each question carefully. Fill in the bubble next to your answer choice for each question.

1 Sara has 36 pumpkin seeds. If the directions on the seed package advise her to plant 3 seeds in each hole, how many holes will she have to dig?

- ○ **A** 3
- ○ **B** 9
- ○ **C** 12
- ○ **D** 15

2 A stone walkway is 108 inches long. If each paving stone is 8 inches long, what is the greatest number of paving stones that can fit lengthwise in the walkway?

- ○ **F** 13
- ○ **G** 11
- ○ **H** 10
- ○ **J** 8

3 Charlotte made 84 flyers to give to the residents of her building. She plans to give each apartment one flyer. If there are 6 apartments on each floor, how many floors are there in her building?

- ○ **A** 8
- ○ **B** 14
- ○ **C** 17
- ○ **D** 21

4 Malcolm feeds his cat 12 ounces of food a day. How many ounces of food will Malcolm need for 9 days?

- ○ **F** 72
- ○ **G** 84
- ○ **H** 96
- ○ **J** 108

Directions: Read each question carefully and write your answer on the line provided.

5 Ms. Bennett bought sets of dishes for her home. Each set of dishes includes a dinner plate, a salad plate, a bowl, a cup, and a saucer. If she bought 40 pieces, how many sets did she buy?

Show your work.

Answer ____________ sets

6 Stephanie's class is going on a field trip. Each student brings in $7 to pay for the trip. If there are 24 people in Stephanie's class, how much did they pay altogether?

Show your work.

Answer $ ____________

Mile 12: Problem Solving

Some problems take more than one step to solve. When you come across problems with multiple steps, first read the problem carefully. Identify any key words that will tell you how to solve the problem. Then, decide which operations you need to use and solve.

Directions: Read each clue carefully. Solve the crossword puzzle by writing out the answer to each clue in the appropriate box.

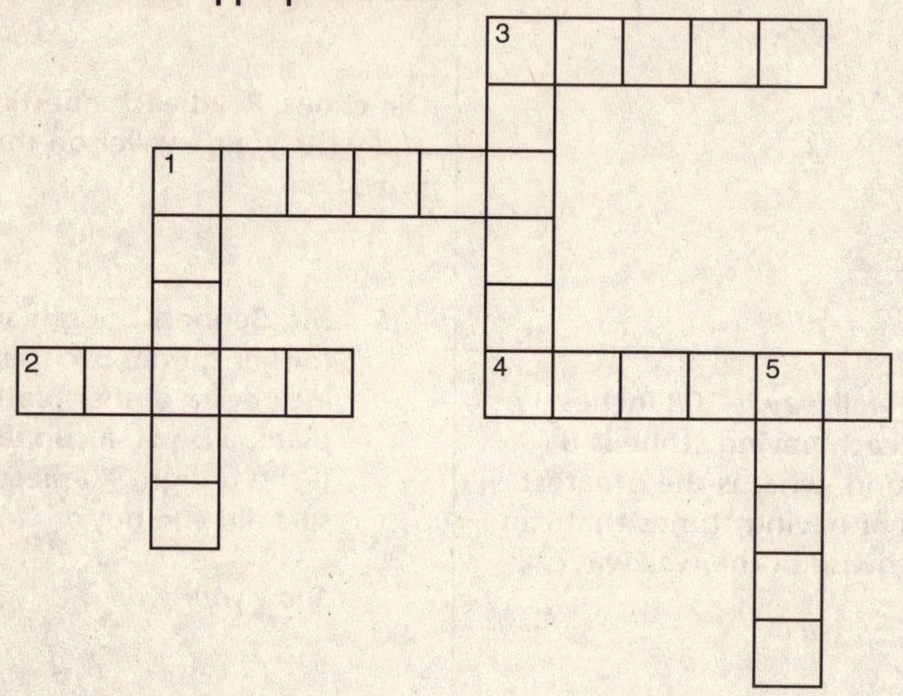

Across

1. Phillipa had 9 marbles. She bought 7 more. She gave her friend 4. How many marbles did Phillipa end up with?
2. Jeremy will be 11 on July 21, 2005. How old will Jeremy be on July 21, 2034?
3. Six questions on a math test were worth 15 points each. One question was worth 10 points. If all the questions and the extra credit question were answered correctly, it was possible to earn a score of 108. How many points was the extra credit question worth?
4. Rashid and his grandfather share the same birthday, March 6. Rashid was born in 1992. How old will Rashid's grandfather be on March 6, 2011, if he is 71 years older than Rashid?

Down

1. A window-washing company charged $25 per job and an additional $5 per window. If they were paid $175 for one job, how many windows did they wash?
3. Lucy saw 4 blue jays on her hike. She saw 3 times as many sparrows as blue jays. She saw 1 more sparrow than robins. How many robins did Lucy see on her walk?
5. Troy had $21. He bought 9 notebooks for $2 each. How much money did he have left over?

Directions: Read each question carefully. Fill in the bubble next to your answer choice for each question.

1 Edward puts his model cars into cases. Each case holds 16 cars. If he has 52 cars, how many cars will ***not*** fit into a full case?

- ○ **A** 4
- ○ **B** 3
- ○ **C** 2
- ○ **D** 1

2 Candace has 11 nickels. She has 12 more pennies than nickels. She has 7 more dimes than pennies. How many dimes does she have?

- ○ **F** 19
- ○ **G** 23
- ○ **H** 30
- ○ **J** 33

3 Tara earned extra spending money by mowing lawns all summer. She spent $50 of her money on new clothes and $15 on a paint set. She put the remaining $40 into a savings account. How much money did Tara earn mowing lawns?

- ○ **A** $35
- ○ **B** $90
- ○ **C** $105
- ○ **D** $115

4 In Ms. Jones's science class, students observed plants growing. Each week, the number of leaves on the plants doubled. After two weeks, there were 40 leaves. How many leaves did the plants have originally?

- ○ **F** 35
- ○ **G** 25
- ○ **H** 20
- ○ **J** 10

Directions: Read each question carefully and write your answer on the line provided.

5 Andy had twice as many red cards as yellow ones. He had 3 less green ones than yellow ones. If he had 6 green cards, how many red cards did he have?

Show your work.

Answer ____________ red cards

6 Rose's little sister is 36 inches tall. Her father is twice the height of Rose's sister. Rose is 19 inches shorter than her father. How tall is Rose?

Show your work.

Answer ____________ inches

MILE 13: DECIMALS

A **decimal** is a number that shows part of a whole. It contains a decimal point and numbers after it. (Sometimes whole numbers come before the decimal point.) The numbers after the decimal point hold place values. For example, in the decimal 1.25, 1 is in the ones place, 2 is in the *tenths* place, and 5 is in the *hundredths* place. Sometimes decimals can be shown with pictures.

The figure at right shows the decimal 1.25.

Each large square represents one whole and is made up of one hundred smaller squares. The 1 in 1.25 is shown by the square that has all 100 smaller squares shaded gray. The 0.25 is shown by the 25 gray squares in the second large square.

Directions: Color the parts of each figure to match the decimal below it. The first one has been done for you.

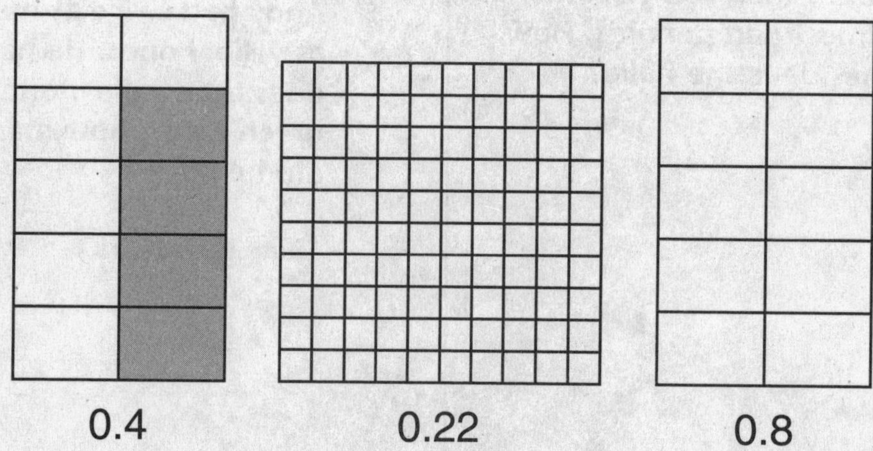

Now write the decimal represented by the shaded part of each of the following figures. The first one has been done for you.

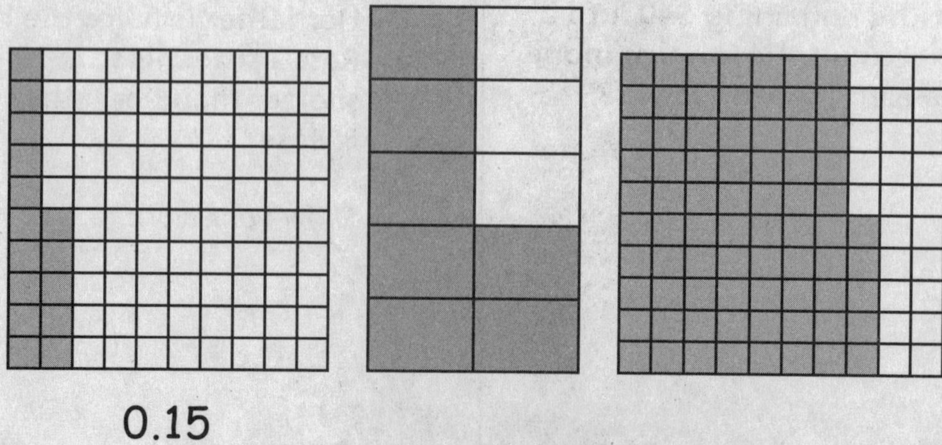

__________ __________ __________

Directions: Read each question carefully. Fill in the bubble next to your answer choice for each question.

1 **It snowed 2.6 inches yesterday. Which arrow on the number line points to the amount it snowed?**

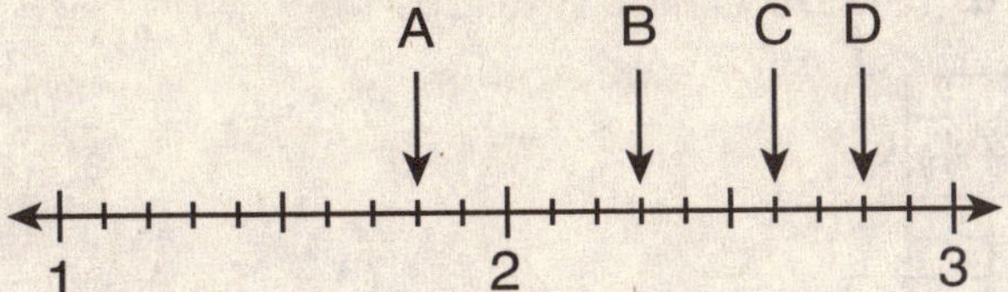

○ **A** A

○ **B** B

○ **C** C

○ **D** D

2 **Which set of numbers are in order from least to greatest?**

○ **F** 3.45, 3.57, 3.84, 3.76

○ **G** 3.57, 3.76, 3.45, 3.84

○ **H** 3.84, 3.76, 3.57, 3.45

○ **J** 3.45, 3.57, 3.76, 3.84

3 **Which of the following represents the decimal model below?**

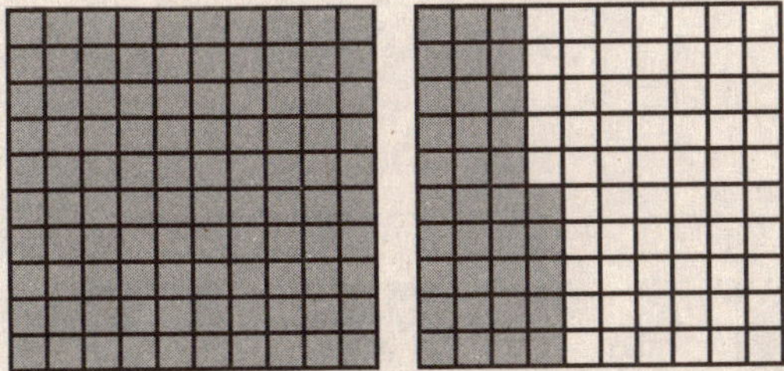

○ **A** 1.25

○ **B** 1.35

○ **C** 1.45

○ **D** 1.55

4 **Which of the following represents the decimal model below?**

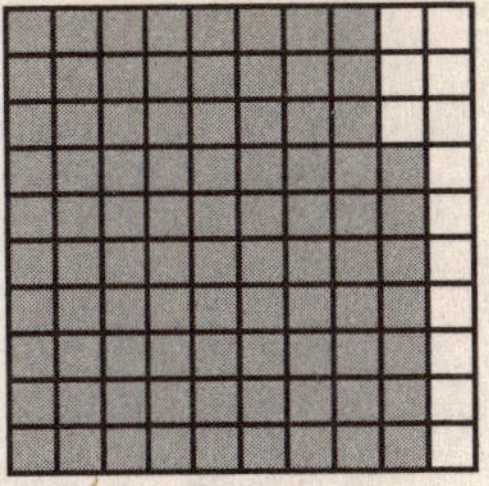

○ **F** 0.87

○ **G** 0.90

○ **H** 0.94

○ **J** 0.99

Directions: Read the question carefully and write your answer in the grid below.

5 **Fill in the amount of squares below that represent the decimal 0.17.**

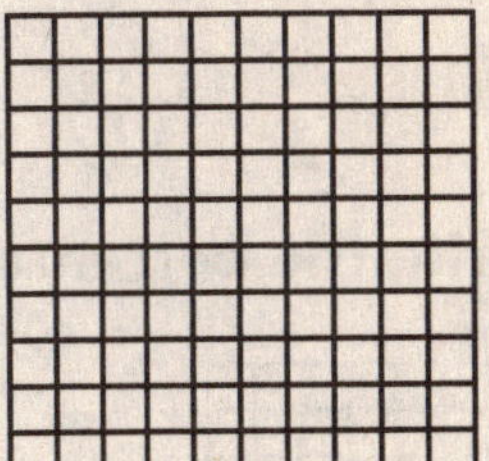

Most errors on decimal problems involve the digit 0. It's easy to put in an extra 0 or to leave a 0 out in a decimal. If you're careful with your zeros, you should have an easier time with decimal problems.

MILE 14: PERCENTAGES

A **percentage** is a way to measure part of a whole. **Percent** means "out of 100." If you divide a whole into 100 parts, you can think of percentages as the part out of 100.

The figure below shows a square divided into 100 equal-sized portions. The square has 70 portions shaded gray. The square is 70% shaded. It is 30% not shaded.

You can also find percentages as the portion of something out of 10. You just multiply it by 10 to get the percent. The figure below shows a square divided into 10 equal-sized portions. The square has 7 portions shaded gray. The square is 70% shaded. Notice that this is similar to the square above with 70 out of 100 portions shaded.

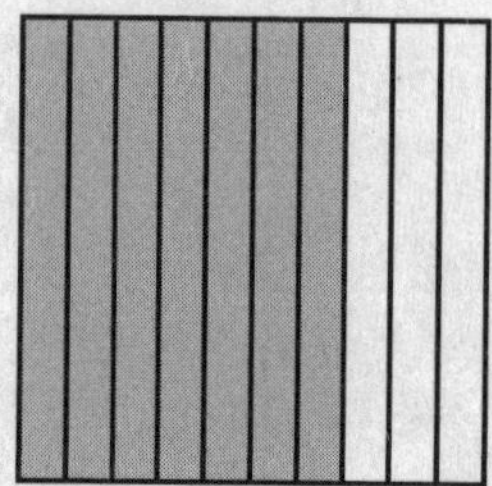

Directions: Write the percentage of the shaded portion of each diagram.

1. 

   ____

2. 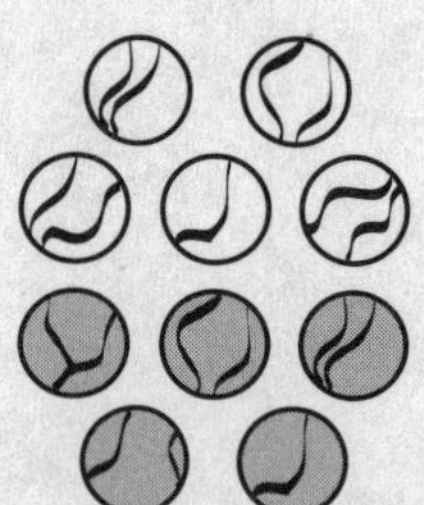

   ____

3. 

   ____

4. 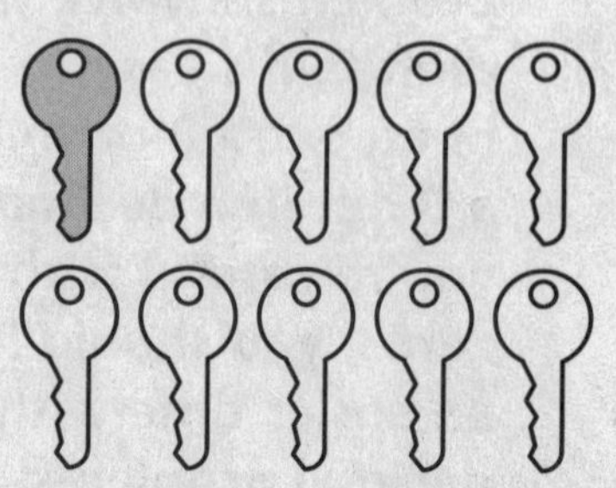

   ____

5. 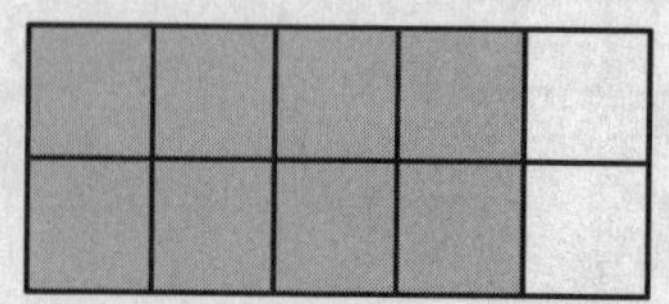

   ____

6. 

   ____

Directions: Read each question carefully. Fill in the bubble next to your answer choice for each question.

1 **Which of the following percentages is the shaded part of the grid?**

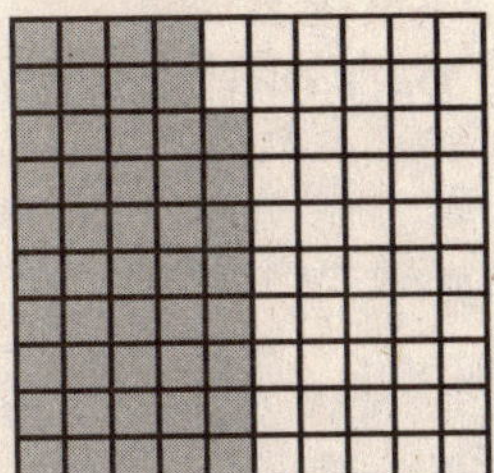

- ○ **A** 48%
- ○ **B** 49%
- ○ **C** 50%
- ○ **D** 52%

2 **Laslo took a math test and got 57 correct out of 100. What percentage did he get right?**

- ○ **F** 4.3%
- ○ **G** 43%
- ○ **H** 57%
- ○ **J** 63%

3 **Liu had 100 pennies. She dropped 20 of her pennies. What percentage of the pennies does Liu still have?**

- ○ **A** 2%
- ○ **B** 20%
- ○ **C** 8%
- ○ **D** 80%

4 **What percentage of the following diagram is *not* shaded?**

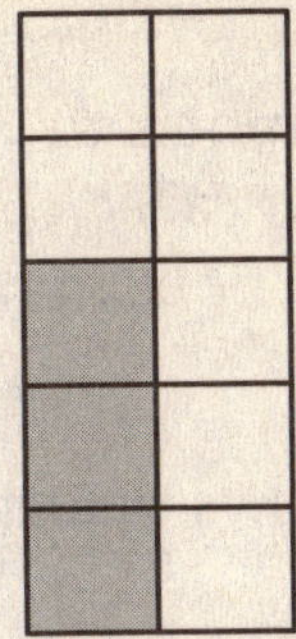

- ○ **F** 30%
- ○ **G** 50%
- ○ **H** 60%
- ○ **J** 70%

Directions: Read the question carefully and write your answer on the line provided.

5 **Put the following percentages in order from least to greatest.**

54% 67% 53% 62% 60%

Answer ______________________

Mile 15: Writing Fractions

You need two numbers to write a **fraction**—one on the top and one on the bottom. The top number is called the **numerator** and shows how many parts of the whole you are counting. The bottom number is the **denominator** and shows how many parts are in the whole.

The fraction $\frac{1}{2}$ shows that there are two parts of a whole. One part of the whole is being counted.

Example:

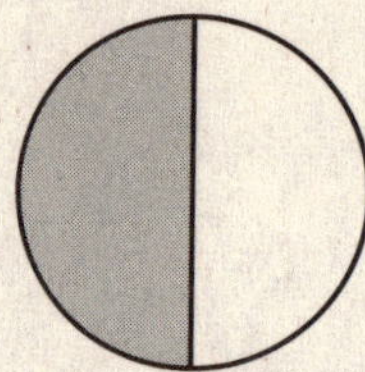

The circle above is split into two parts. One part of the circle is shaded. So $\frac{1}{2}$ of the circle is shaded.

Directions: Look at this picture of a fence. Then read and answer the questions that follow.

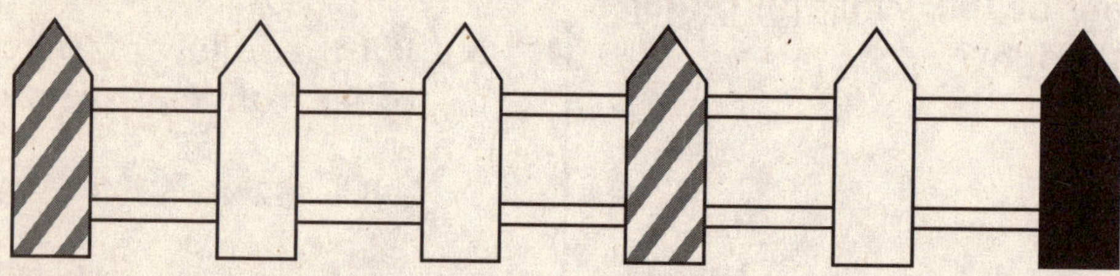

1. **How many fence posts are there?** ____________

2. **What fraction shows the number of fence posts with stripes?** ____________

3. **What fraction shows the number of white fence posts?** ____________

4. **What fraction of the fence posts are black?** ____________

Directions: Read each question carefully. Fill in the bubble next to your answer choice for each question.

1 What fraction of the square is shaded?

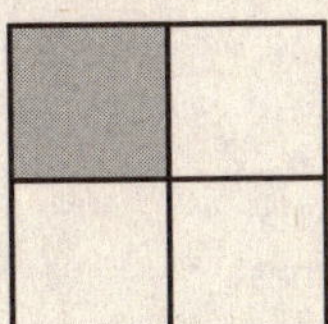

○ **A** $\frac{1}{5}$

● **B** $\frac{1}{4}$

○ **C** $\frac{1}{3}$

○ **D** $\frac{1}{2}$

2 What fraction shows the number of bottles filled with soda?

○ **F** $\frac{1}{6}$

○ **G** $\frac{3}{6}$

○ **H** $\frac{4}{6}$

● **J** $\frac{5}{6}$

3 What fraction of the diagram is shaded?

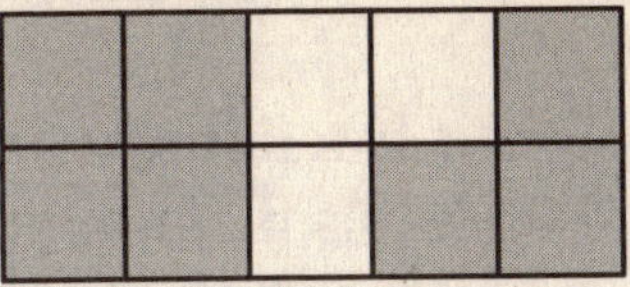

○ **A** $\frac{4}{10}$

○ **B** $\frac{5}{10}$

● **C** $\frac{7}{10}$

○ **D** $\frac{8}{10}$

Directions: Read each question carefully and write your answer on the line provided.

4 What fraction of the octagon is shaded?

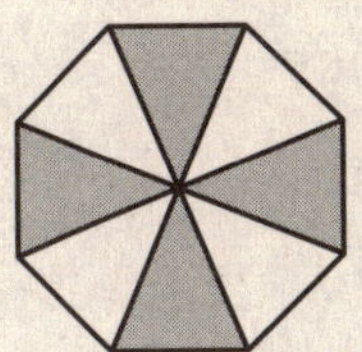

Answer $\frac{4}{8}$ or $\frac{1}{2}$

5 What fraction of the figure below is shaded?

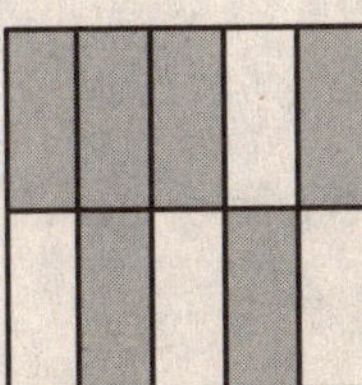

Answer $\frac{6}{10}$

Mile 16: Comparing Fractions

When **comparing fractions,** you see how numbers relate or compare to each other. You can compare fractions by finding equivalent fractions with the same denominator. You can also use a number line.

Which is larger, $\frac{5}{6}$ or $\frac{2}{3}$?

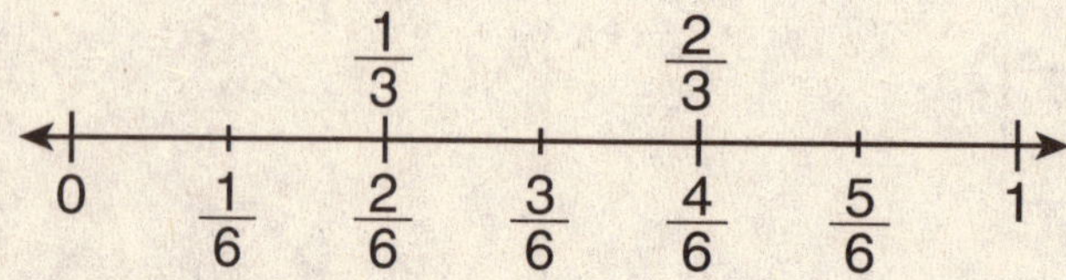

You can see that $\frac{5}{6}$ is the larger fraction because it is farther to the right on the number line. You can also see that the fraction $\frac{2}{3}$ is equivalent to $\frac{4}{6}$, which is less than $\frac{5}{6}$.

Directions: Compare the two fractions in each question and circle the larger fraction. Each fraction stands for a word. You will use the words to answer the question at the bottom of the page.

1. $\frac{1}{10}$ **The** $\frac{1}{5}$ **A**

2. $\frac{1}{6}$ **Money** $\frac{1}{4}$ **Penny**

3. $\frac{4}{10}$ **Saved** $\frac{3}{10}$ **Stored**

4. $\frac{1}{3}$ **May** $\frac{1}{2}$ **Is**

5. $\frac{1}{3}$ **Earn** $\frac{2}{3}$ **A**

6. $\frac{5}{6}$ **Penny** $\frac{6}{12}$ **Plenty**

7. $\frac{3}{8}$ **Bank** $\frac{7}{8}$ **Earned**

Answer the following question by writing the word that goes with each fraction in order from 1 to 7.

► **What is Benjamin Franklin's famous saying about money?**

A penny saved is a penny earned.

Directions: Read each question carefully. Fill in the bubble next to your answer choice for each question.

1 Which of the following is larger than $\frac{2}{3}$?

- A $\frac{2}{5}$
- B $\frac{1}{2}$
- C $\frac{4}{6}$
- D $\frac{6}{8}$

2 Which of the following is in the correct order from least to greatest?

- F $\frac{1}{6}, \frac{1}{10}, \frac{1}{8}$
- G $\frac{1}{8}, \frac{1}{6}, \frac{1}{10}$
- H $\frac{1}{6}, \frac{1}{8}, \frac{1}{10}$
- J $\frac{1}{10}, \frac{1}{8}, \frac{1}{6}$

3 Which of the following is in the correct order from least to greatest?

- A $\frac{2}{7}, \frac{2}{11}, \frac{2}{5}$
- B $\frac{2}{11}, \frac{2}{7}, \frac{2}{5}$
- C $\frac{2}{5}, \frac{2}{11}, \frac{2}{7}$
- D $\frac{2}{5}, \frac{2}{7}, \frac{2}{11}$

Directions: Read each question carefully and write your answer on the line or in the box provided.

4 Put the following fractions in order from greatest to least.

$\frac{1}{8}, \frac{5}{8}, \frac{2}{8}, \frac{3}{8}$

Answer $\frac{5}{8}$ $\frac{3}{8}$ $\frac{2}{8}$ $\frac{1}{8}$

5 Put the following fractions in order from least to greatest.

$\frac{3}{8}, \frac{3}{5}, \frac{3}{10}, \frac{3}{4}$

Answer $\frac{3}{10}$ $\frac{3}{8}$ $\frac{3}{5}$ $\frac{3}{4}$

6 Write >, <, or = in each box.

$\frac{1}{5}$ [<] $\frac{1}{4}$ $\frac{3}{6}$ [=] $\frac{1}{2}$ $\frac{4}{6}$ [>] $\frac{4}{8}$

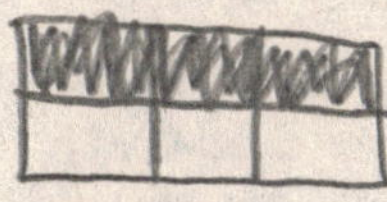

MILE 17: PROBLEM SOLVING WITH FRACTIONS

Sometimes you will need to solve problems by finding a fraction of a whole number. You can use counters, like the ones cut out of the tool sheet on page 7, or draw a diagram to solve these kinds of problems.

If you want find $\frac{1}{2}$ of 6, take out or draw 6 counters. Divide the counters into 2 equal groups. Then count the number of objects in one group. There are 3 counters in one group. That means $\frac{1}{2}$ of 6 = 3.

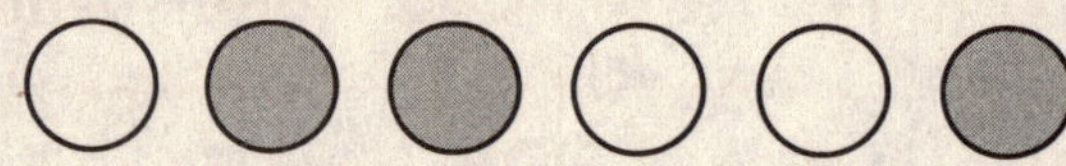

You can also find a fraction of a whole number by multiplying the number by the fraction's numerator and dividing the product by the fraction's denominator. If you want to find $\frac{2}{3}$ of 9, multiply 9 by the numerator, 2, to get 18. Then divide 18 by the denominator, 3, which gives you 6. That means $\frac{2}{3}$ of 9 = 6.

Directions: You may use your counters to find the answers to the questions below. When you are done, write the letter to the right of each problem in the box below that contains the problem's correct answer. The letters will spell out the name of the longest river in the United States.

1. $\frac{1}{4}$ of 20 = ______ O

2. $\frac{1}{3}$ of 3 = ______ M

3. $\frac{1}{2}$ of 14 = ______ R

4. $\frac{3}{4}$ of 8 = ______ U

5. $\frac{2}{3}$ of 12 = ______ I

6. $\frac{1}{6}$ of 12 = ______ I

7. $\frac{2}{5}$ of 10 = ______ S

8. $\frac{1}{5}$ of 15 = ______ S

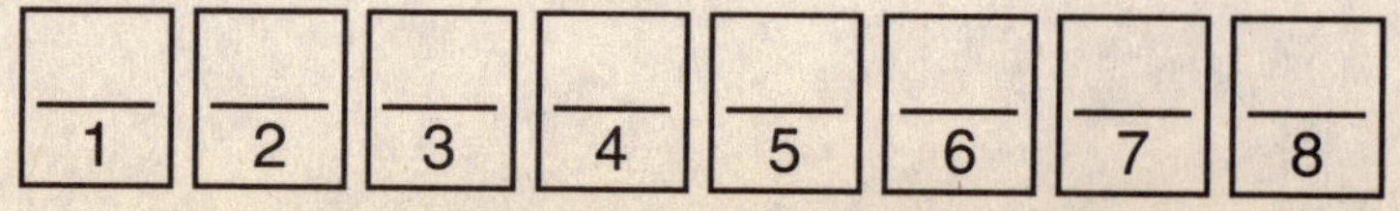

Directions: Read each question carefully. Solve these multiple-step word problems using your counters. Fill in the bubble next to your answer choice for each question.

1 Steven had 12 pencils. He gave $\frac{1}{2}$ of them to Susan. She gave $\frac{1}{2}$ of her pencils to Karen. How many pencils did Karen get?

○ **A** 2

○ **B** 3

○ **C** 4

○ **D** 6

2 The All-Sports Store had a baseball department. They had $\frac{3}{4}$ as many bats as baseball gloves. There were $\frac{1}{3}$ as many catchers' masks as bats. If there were 16 baseball gloves, how many catchers' masks were there?

○ **F** 4

○ **G** 8

○ **H** 12

○ **J** 16

3 Stacy made 3 dozen oatmeal cookies. She ate $\frac{1}{6}$ of them. How many did she have left?

○ **A** 12

○ **B** 24

○ **C** 30

○ **D** 32

Directions: Read each question carefully and write your answer on the line provided.

4 Kiernan and Lois started growing tomato plants from seeds. Kiernan planted 18 plants in his yard, but $\frac{2}{3}$ of them died. Lois planted 12 plants in her yard, but $\frac{1}{6}$ of them died. Who had the most plants left, Kiernan or Lois? How many more plants did Kiernan or Lois have?

Show your work.

Answer ____________

5 Alfred's fish tank has 15 fish in it. $\frac{2}{5}$ of them are guppies and $\frac{1}{2}$ of the guppies are males. How many female guppies are there?

Show your work.

Answer ____________ female guppies

6 Cleo has a drawer with differently colored socks. There are 24 socks in all, $\frac{1}{3}$ of the socks are black, and $\frac{1}{4}$ of the black socks have holes in them. How many black socks have holes in them?

Show your work.

Answer ____________ black socks

Mile 18: Estimation

You use **estimation** to quickly find a number that is close to the correct answer you are looking for. Estimating helps you quickly calculate *about* what the solution will be, so you can check your work or eliminate wrong answer choices on a test. Estimating uses rounded numbers for easier calculations.

Directions: Round each of the following numbers to the nearest ten. Round each of the decimals to the nearest tenth and write it as a fraction. Then find the rounded numbers in the code box below and cross out those boxes.

1. 54 __________

2. 88 __________

3. 243 __________

4. 171 __________

5. 497 __________

6. 0.47 $\frac{\square}{\square}$

7. 0.77 $\frac{\square}{\square}$

8. 0.32 $\frac{\square}{\square}$

Directions: Rewrite each problem with rounded numbers and perform the operation. Find the estimate in the code box below and cross out those boxes.

9. 187 + 83   + ______

10. 98 × 11   × ______

11. 9)62   )______

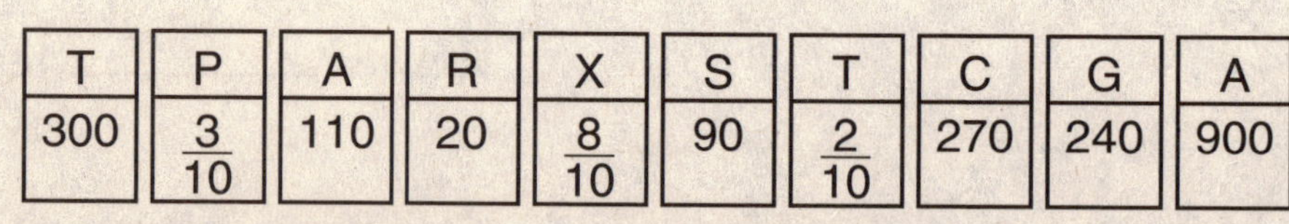

T	P	A	R	X	S	T	C	G	A
300	$\frac{3}{10}$	110	20	$\frac{8}{10}$	90	$\frac{2}{10}$	270	240	900

B	A	F	V	U	M	Z	N	D	L
170	5	50	6	400	1000	500	230	$\frac{5}{10}$	550

Take the letters that are not used in the code box and list them in the boxes below according to their numbers from *least* to *greatest*.

▶ **I am a hairy, eight-legged creature that lives in the desert. What am I?**

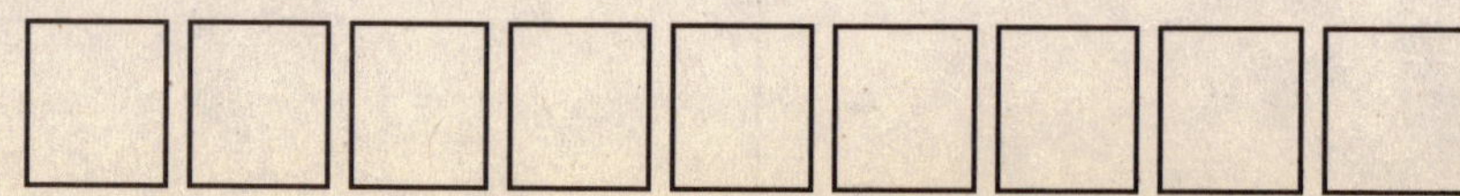

Directions: Read each question carefully. Fill in the bubble next to your answer choice for each question.

1 **Mr. Perry made pies that were sold at a local bakery. He made 17 pies in one week, 11 pies in another week, and 21 pies last week. About how many pies did he make in all?**

○ **A** 40

○ **B** 50

○ **C** 60

○ **D** 70

2 **Last winter, Marie went cross-country skiing every weekend. If she skied a total of 307 miles in 10 weeks, about how many miles did she ski each week?**

○ **F** 10

○ **G** 20

○ **H** 30

○ **J** 40

3 **The table below shows the weight gain of Elaine's younger sister soon after she was born. In which month did Elaine's sister gain closest to $1\frac{1}{4}$ pounds?**

Month	Weight Gain
May	1.3 lbs.
June	2.8 lbs.
July	1.9 lbs.
August	1.6 lbs.

○ **A** May

○ **B** June

○ **C** July

○ **D** August

Directions: Read each question carefully and write your answer on the line provided.

4 **Tyrell and his friends grew different amounts during the past year. Which of them grew closest to $7\frac{3}{4}$ centimeters in the past year?**

Name	Growth
Tyrell	7.4 cm
Michael	7.8 cm
Lavar	6.9 cm
Shane	6.8 cm

Show your work.

Answer __________

5 **Ms. Mary mixes juice drink for her preschool class each day. If she mixed a total of 122 ounces of juice in 5 days, about how much juice did she make each day?**

Show your work.

Answer __________ **ounces**

MILE 19: ESTIMATING MEASUREMENT

When you **estimate a measurement**, you are making a guess about how much something actually measures. You can estimate the length, weight, or volume of everyday objects.

Directions: Read the chart below and fill in the missing units of measurement.

Length

1 foot = ______ inches
1 yard = ______ feet
1 meter = ______ centimeters

Weight

1 pound = ______ ounces
1 kilogram = ______ grams

Liquid Volume

1 pint = ______ cups
1 quart = ______ pints
1 gallon = ______ quarts
1 liter = ______ milliliters

Read each statement carefully. Decide whether each statement is true or false. Use the measurement chart above to help you. (You can check your answers on page 91.) Circle the T if the statement is true. Circle the F if the statement is false.

1. T F The width of a kite is easily measured using inches.

2. T F The amount of juice in a glass is best measured using gallons.

3. T F The weight of a computer is best measured in grams.

4. T F The length of a dog's tail should be measured in yards.

5. T F The amount of liquid in a pot of water is most accurately measured in quarts.

Directions: Read each question carefully. Fill in the bubble next to your answer choice for each question.

1 **Which measurement is the best estimate for the length of a school bus?**

○ **A** 13 inches

● **B** 13 feet

○ **C** 13 yards

○ **D** 13 miles

2 **Which measurement is the best estimate for the weight of a textbook?**

○ **F** 3 grams

● **G** 3 pounds

○ **H** 3 ounces

○ **J** 3 tons

3 **Which measurement is the best estimate for the amount of milk that can fit in a mug?**

● **A** 1 cup

○ **B** 1 quart

○ **C** 1 gallon

○ **D** 1 liter

4 **Sarah walked across the street to her friend Larry's house. Which of the following would be the best way to measure the distance she walked?**

○ **F** centimeters

○ **G** inches

● **H** feet

○ **J** miles

5 **Which of the following would weigh about 2,000 kilograms?**

○ **A** a school desk

● **B** an 18-wheel truck

○ **C** a human adult

○ **D** a home computer

6 **Which of the following would hold about 35 gallons of water?**

● **F** a gym swimming pool

○ **G** a goldfish bowl

○ **H** a soup pot

○ **J** a bathtub

Some answer choices for questions that ask you to estimate are much too large, and others are much too small. Get rid of them! Then you'll have a better chance of getting the correct answer.

MILE 20: USING A RULER

When you measure an object with a ruler, you are figuring out how long it is, or its length. Objects can be measured in centimeters, inches, feet, yards, and meters. The New York Grade 4 Math test will tell you when to use a ruler to answer a question.

Directions: Get some practice measuring the lengths of objects. Use a ruler to measure the length of each object shown below in centimeters and inches. Write the measurement on the line provided near the object.

1.

_____ centimeters
_____ inches

2.

_____ centimeters
_____ inches

3.

marker+
PURPLE

_____ centimeters
_____ inches

4.

_____ centimeters
_____ inches

5.

_____ centimeters
_____ inches

Directions: Read each question carefully. Use your ruler to answer the following questions. Fill in the bubble next to your answer choice for each question.

1 **What is the length of the turtle?**

- ○ **A** 4 centimeters
- ○ **B** 5 centimeters
- ○ **C** 6 centimeters
- ○ **D** 7 centimeters

2 **Kelly and Jake cut out paper snowflakes to decorate their windows. Kelly's snowflake is $8\frac{1}{2}$ centimeters wider than Jake's snowflake. How wide is Kelly's snowflake?**

Jake's Snowflake

- ○ **F** $4\frac{1}{2}$ centimeters
- ○ **G** $11\frac{1}{2}$ centimeters
- ○ **H** $12\frac{1}{2}$ centimeters
- ○ **J** 13 centimeters

Directions: Read the question carefully and write your answers on the lines provided.

3 **Use your ruler to measure the lengths of the figure below. Write the measurements on the lines provided.**

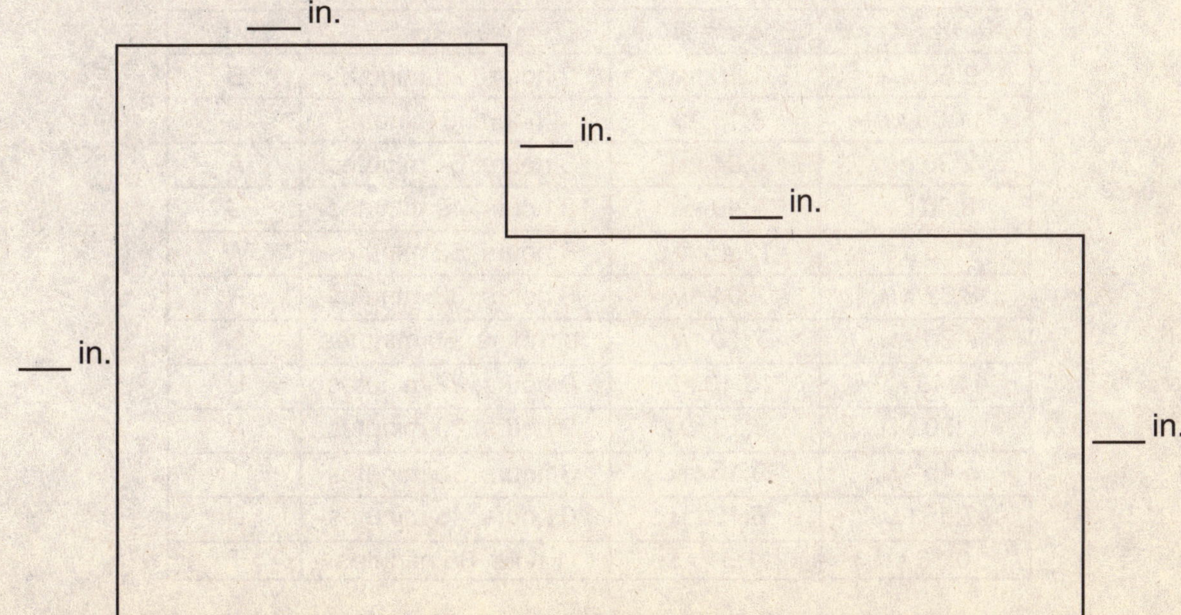

MILE 21: TIME

Telling time is important for getting to school before class starts, making a matinee movie, or planning a schedule. For most of these activities, you'll want to figure out how much time it takes to do something, like brushing your teeth or walking to school, so you'll be able to get to a certain event on time. The time that passes is called elapsed time.

Be careful when calculating elapsed time. Suppose you want to find out how much time passed between 2:30 P.M. and 3:25 P.M. You cannot just subtract 230 from 325. If you look at a clock and count the minutes between the two times, you will discover that there are 55 minutes between 2:30 P.M. and 3:25 P.M.

Now try to calculate the elapsed time between 10:55 A.M. and 1:24 P.M. First, count all the complete hours between the two times: 11:00 A.M. to 12:00 P.M. and 12:00 P.M. to 1:00 P.M. There are 2 complete hours. Next, count the number of minutes from 10:55 A.M. until the end of the hour (11:00 A.M.). There are 5 minutes. Next, count the number of minutes from 1:00 P.M. to 1:24 P.M. There are 24 minutes. Add 5 and 24 to get 29 minutes. The total elapsed time between 10:55 A.M. and 1:24 P.M. is 2 hours and 29 minutes.

Directions: Calculate the amount of time that passes between the starting time and ending time listed in each row of the table below. Look at the value in the third column, marked "Time Passed?" If the value is correct, leave the row as is. If the value is incorrect, cross out the letter in the fourth column. When you are done, the remaining letters will spell out the answer to the riddle.

► **Question: What did the lawyer wear to court? Answer: A ___________**

Starting Time	Ending Time	Time Passed?	Letter
9:55 A.M.	1:10 P.M.	2 hours, 25 minutes	B
11:09 A.M.	3:15 P.M.	4 hours, 6 minutes	L
2:30 P.M.	8:04 P.M.	5 hours, 34 minutes	A
3:00 P.M.	7:49 P.M.	3 hours, 49 minutes	C
7:55 A.M.	12:48 P.M.	4 hours, 53 minutes	W
8:32 A.M.	2:04 P.M.	8 hours, 32 minutes	K
7:20 A.M.	7:10 P.M.	11 hours, 50 minutes	S
11:48 A.M.	12:10 P.M.	0 hours, 22 minutes	U
6:50 A.M.	9:10 A.M.	3 hours, 20 minutes	N
5:45 A.M.	9:15 A.M.	3 hours, 30 minutes	I
7:45 P.M.	8:15 P.M.	0 hours, 45 minutes	D
10:16 A.M.	11:51 A.M.	1 hour, 35 minutes	T

Directions: Read each question carefully. Fill in the bubble next to your answer choice for each question.

1 A play began at 7:20 P.M. and ended at 9:00 P.M. How long did the play last?

- ● **A** 1 hour, 40 minutes
- ○ **B** 1 hour, 50 minutes
- ○ **C** 2 hours, 16 minutes
- ○ **D** 2 hours, 24 minutes

2 Soccer practice will be over in 1 hour and 35 minutes. It is now 4:40 P.M. What time will practice be over?

- ○ **F** 5:15 P.M.
- ○ **G** 5:40 P.M.
- ● **H** 6:15 P.M.
- ○ **J** 6:40 P.M.

3 Marvin puts bread into the oven at 2:30 P.M. It needs to bake for 55 minutes. Based on the clock below, how much longer does the bread need to bake?

- ○ **A** 15 minutes
- ○ **B** 30 minutes
- ● **C** 40 minutes
- ○ **D** 55 minutes

Directions: Read each question carefully and write your answer on the line provided.

4 Ervin rode to the library on his bicycle. He left home at 4:28 P.M. If he returned home in 1 hour and 17 minutes, what time did he get home?

Show your work.

1 hour 17 minutes
4:28 5:28 5:45

Answer 5:45 P.M.

5 The first school bell rang at 8:37 A.M. The last school bell rang 6 hours and 19 minutes later. What time did the last school bell ring?

Show your work.

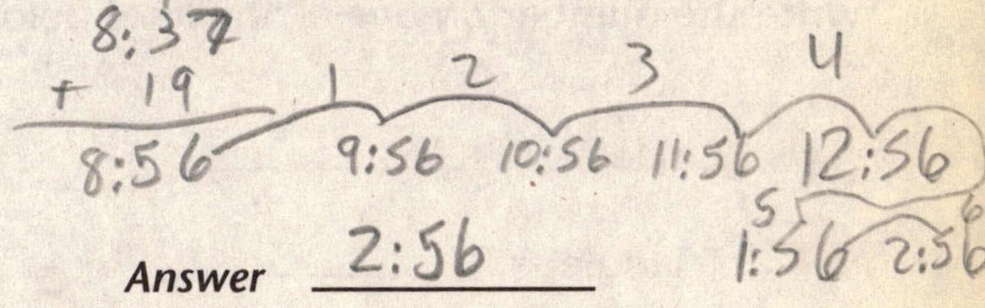

Answer 2:56

Be careful when you figure out how much time has passed between one time and another. You cannot just add and subtract times the way you would whole numbers. That's because there are 60 minutes in an hour.

MILE 22: IDENTIFYING SHAPES

There are many different kinds of **geometric shapes.** The ones that you will see often are *squares, rectangles, triangles,* and *circles.* You might also know the shape of a *parallelogram,* a *rhombus,* a *trapezoid,* and a *hexagon.*

Directions: Unscramble the letters under each shape to find the correct name of the shape.

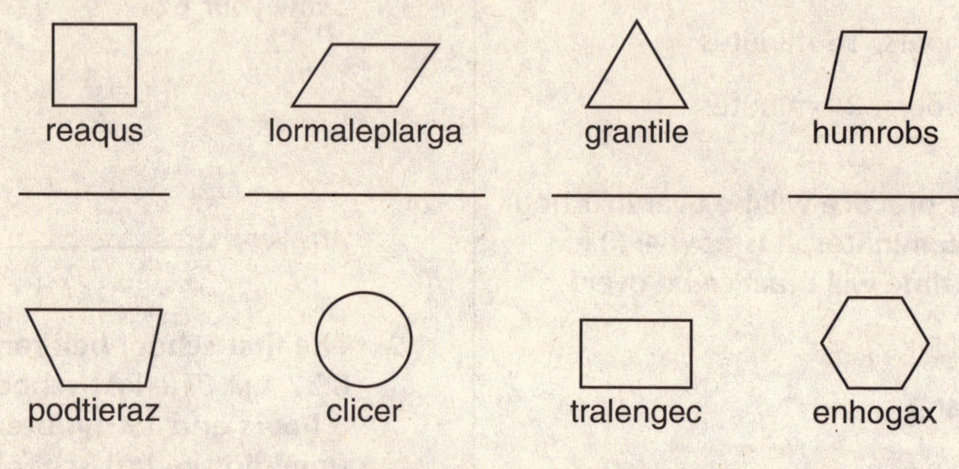

Directions: Look at the shapes in the design below. Identify each numbered shape and write the number on one of the lines below.

Circle = ____________ **Parallelogram =** ____________

Triangle = ____________ **Rectangle =** ____________

Trapezoid = ____________ **Square =** ____________

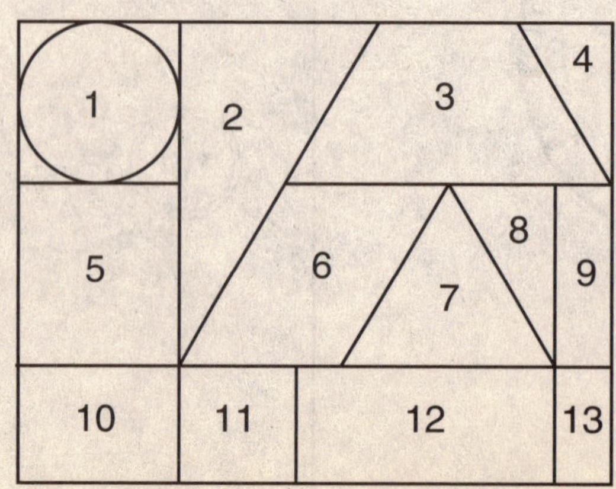

Directions: Read each question carefully. Fill in the bubble next to your answer choice for each question.

1 **Look at the picture of the school lunch. Which of the following shapes is shown most often in the picture?**

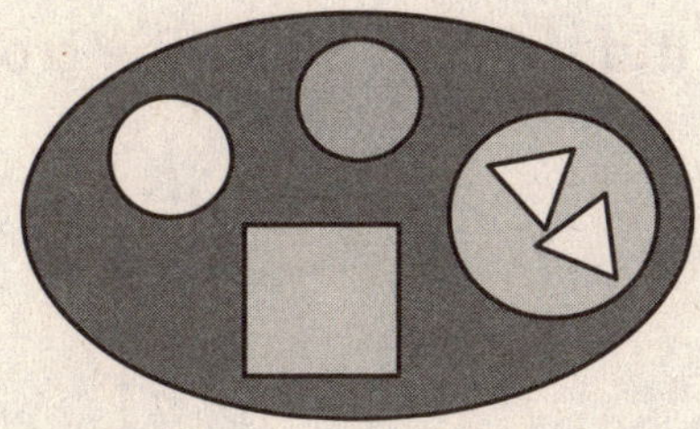

○ **A** triangle

○ **B** square

○ **C** rectangle

○ **D** circle

2 **What shape is not in the painting?**

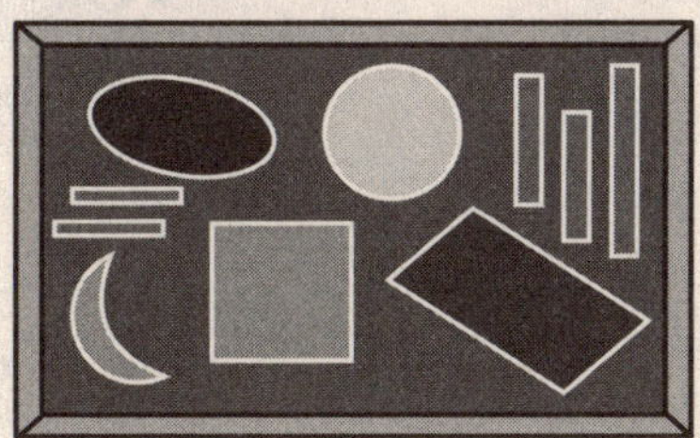

○ **F** square

○ **G** parallelogram

○ **H** triangle

○ **J** circle

3 **The diagram below has five shapes. Which of the following is the correct order of the shapes from top to bottom?**

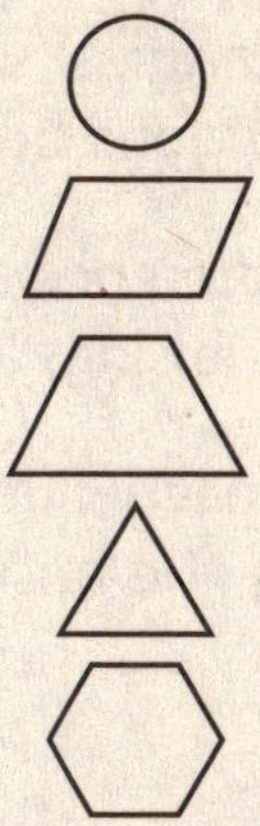

○ **A** circle, parallelogram, hexagon, trapezoid, triangle

○ **B** circle, hexagon, parallelogram, trapezoid, triangle

○ **C** circle, trapezoid, triangle, parallelogram, hexagon

○ **D** circle, parallelogram, trapezoid, triangle, hexagon

Directions: Read the question carefully and write your answer on the line provided.

4 **The diagram below has five shapes. Which shape is in the middle?**

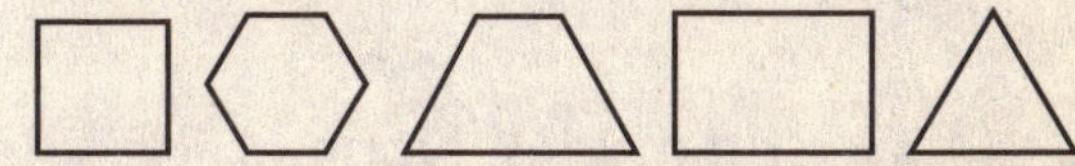

Answer ______________

Mile 23: Properties of Shapes

Different shapes have particular properties that make them unique. Shapes with straight sides, such as rectangles, squares, triangles, and trapezoids, are called *polygons*. Circles are different from polygons because they don't have straight sides.

You can see the difference among shapes based on the number of sides they have. You can also see if any of the sides are parallel or perpendicular.

Directions: Read each clue carefully. Identify the shape it describes. Solve the crossword puzzle by writing the answer in the appropriate box.

Across

1. I have four sides that are all the same length. Each side is perpendicular to the adjacent sides. I have four right angles.

2. I have three sides. None of the sides are parallel.

3. I have six sides.

4. I am round, and I don't have any sides.

Down

5. I have four sides. The opposite sides are parallel and the same length. I have four right angles.

6. I have four sides. I have only one pair of parallel sides.

Directions: Read each question carefully. Fill in the bubble next to your answer choice for each question.

1 Which shape has no parallel lines?

○ **A** 

○ **B**

○ **C** 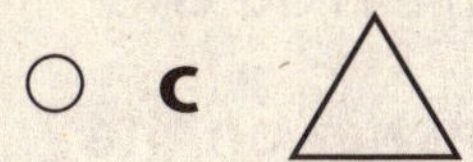

○ **D**

2 Which shape has exactly one pair of parallel lines?

○ **F** 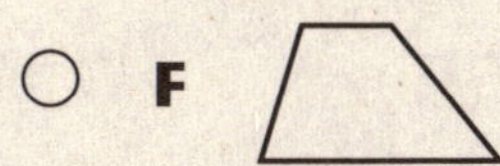

○ **G** 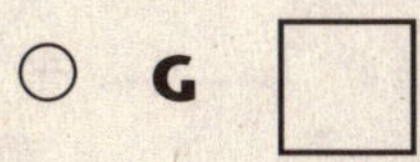

○ **H** 

○ **J**

3 Which shape has exactly two pairs of parallel lines?

○ **A** 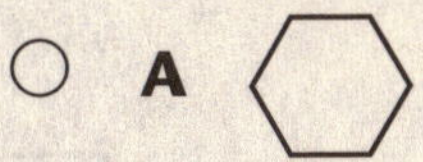

○ **B** 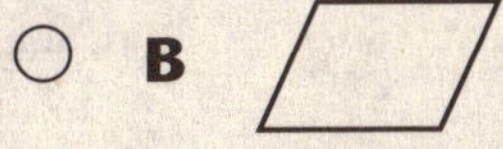

○ **C** 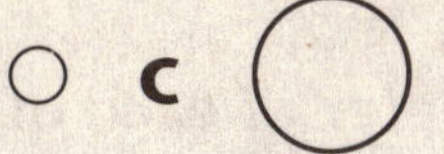

○ **D** 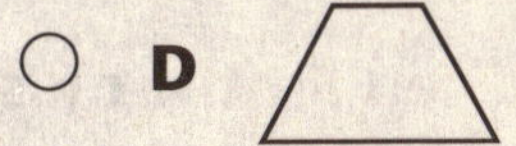

To figure out if a shape has a right angle, place the edge of a piece of paper against the corner of the shape. If the outline of the shape matches the corner of the paper, it's a right angle.

MILE 24: PERIMETER

The distance around a figure is called the **perimeter.** Because it is a distance, perimeter is measured in units of length such as inches, feet, centimeters, or meters. Look at the example below.

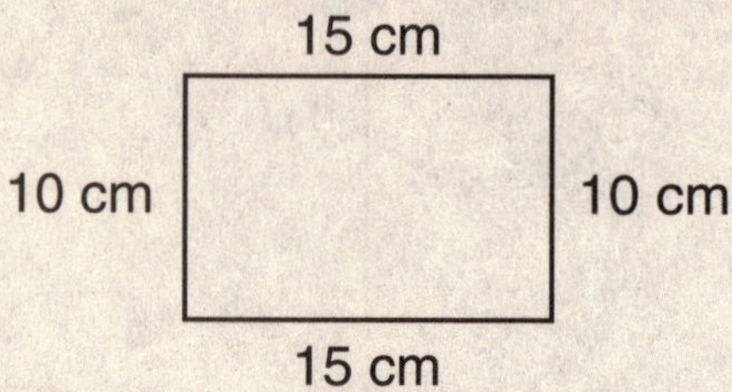

To find the perimeter of the rectangle above, just add up the lengths of all the sides: 10 cm + 10 cm + 15 cm + 15 cm = 50 cm. The perimeter of the rectangle is 50 cm.

Directions: Find the perimeter of each of these six objects. Write your answer on the line provided. Then fill in the letter in the box below that matches the answer the letter goes with. When you fill in all the boxes, you will have the answer to the riddle below.

▶ **What type of bird can lift the most?**

1.

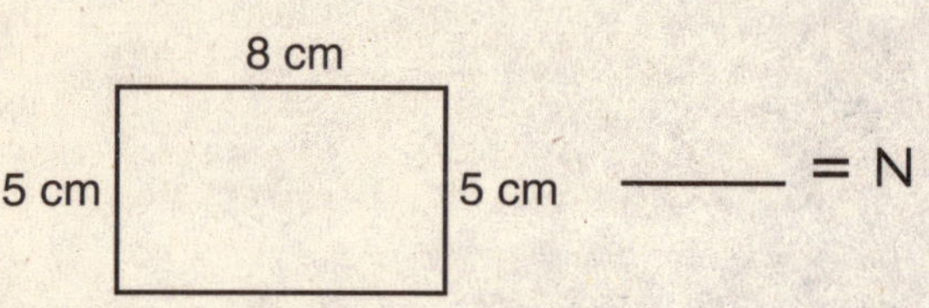

2.

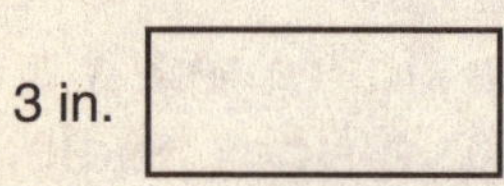

______ = A

3.

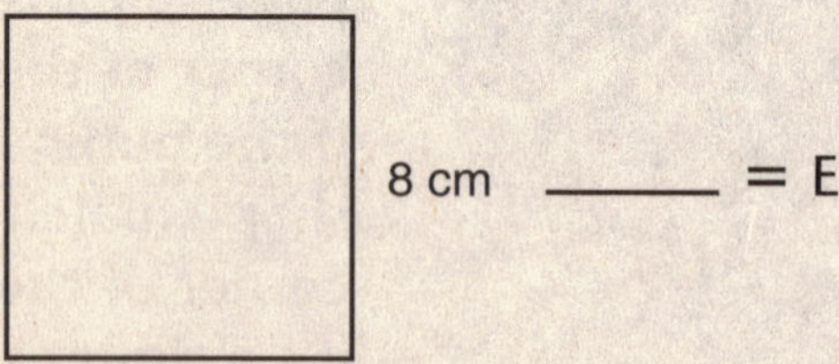

______ = E

4.

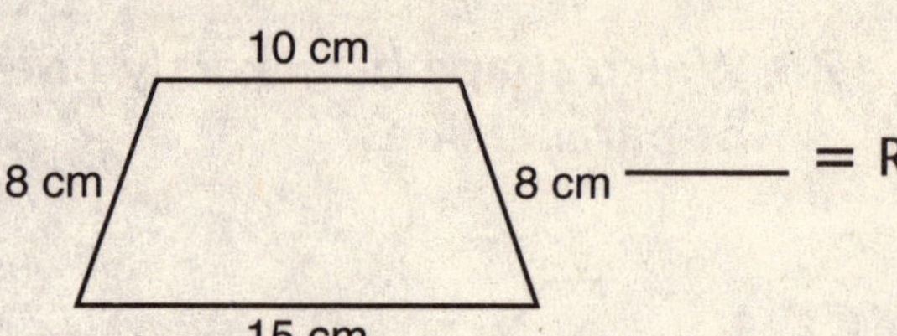

5.

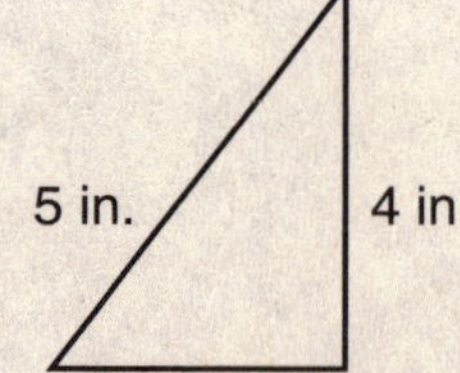

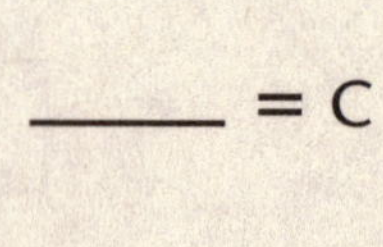

___	___	___	___	___
12 in.	41 cm	20 in.	26 cm	32 cm

Directions: Read each question carefully. Fill in the bubble next to your answer choice for each question.

1 What is the perimeter of this figure?

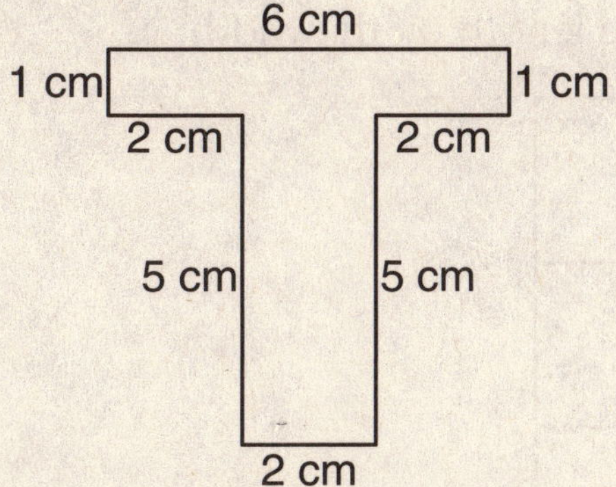

○ **A** 15 centimeters

○ **B** 18 centimeters

○ **C** 20 centimeters

○ **D** 24 centimeters

2 What is the perimeter of this figure?

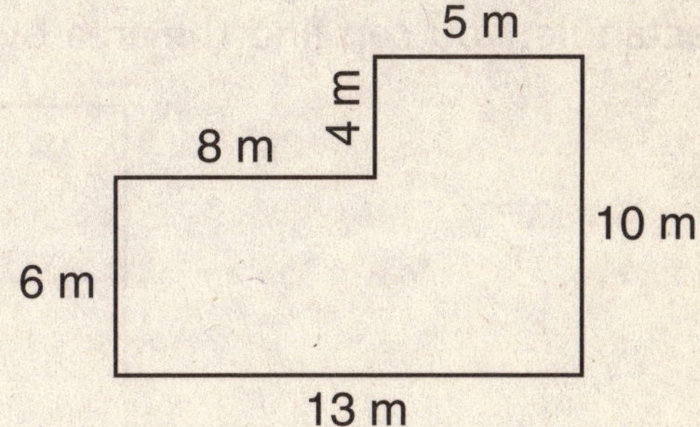

○ **F** 38 meters

○ **G** 43 meters

○ **H** 46 meters

○ **J** 49 meters

Directions: Read the question carefully and write your answer on the line provided.

3 Find the perimeter of the figure below.

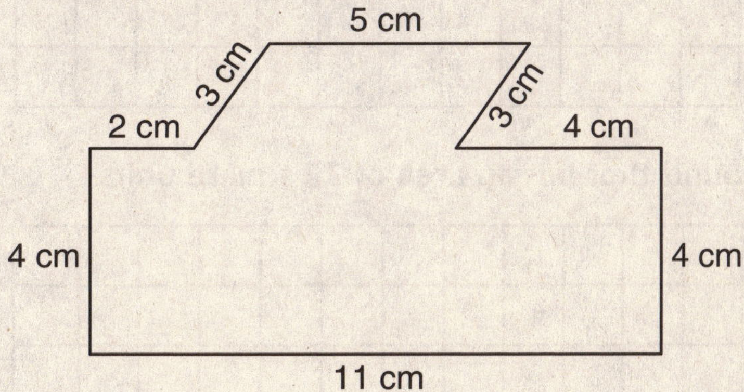

Show your work.

Answer __________ centimeters

You can take a shortcut to find the perimeter of a square. The lengths of all 4 sides of a square are equal. To get the perimeter, just multiply the length of one side by 4.

Mile 25: Area

Area is the amount of space a shape takes up. To find the area, you can count the square units. For example, the figure below has an area of 7 square units. If the shape is a square or a rectangle, you can find the area by multiplying the length by the width.

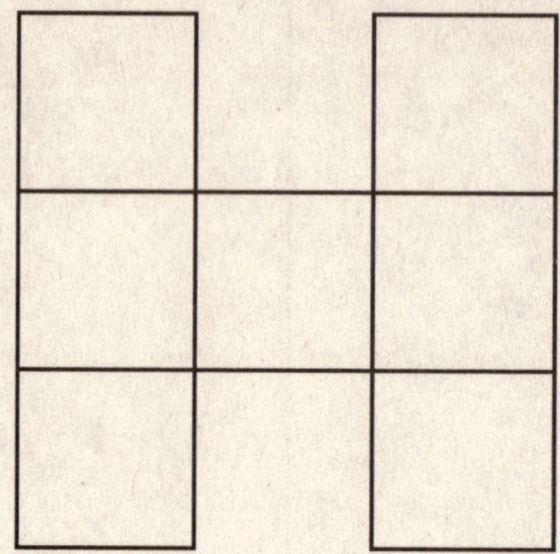

Directions: Draw a picture for each question using the number of square units noted in the question. You can make the pictures as detailed as you want. However, the pictures should not take up more square units than you are allowed.

1. **Draw a board game that has an area of 9 square units.**

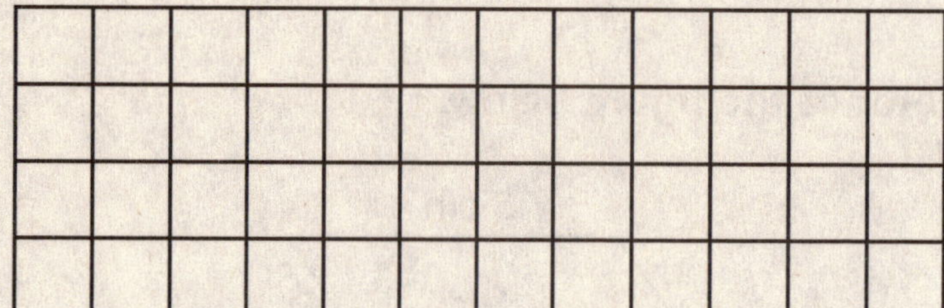

2. **Draw a playground that has an area of 12 square units.**

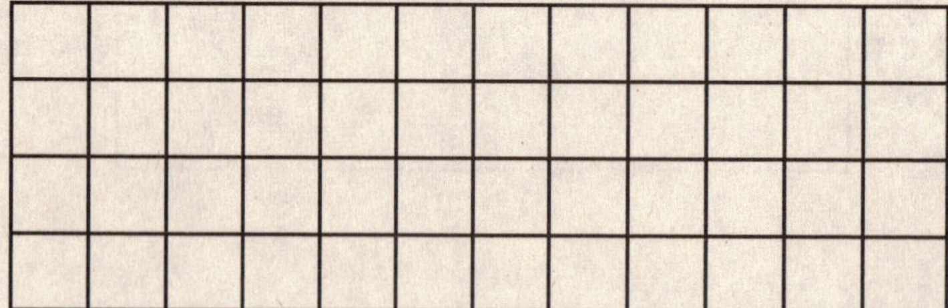

3. **In the space below, draw 10 square units in any pattern you like.**

Directions: Read each question carefully. Fill in the bubble next to your answer choice for each question.

1 **Sierra is making a picture frame. To make a pattern for the frame, she shaded some square units on the piece of paper below.**

How many square units make up her pattern for the frame?

○ **A** 20

○ **B** 22

○ **C** 24

○ **D** 28

2 **Each tile on the floor below has an area of 4 square units. What is the area of the entire floor?**

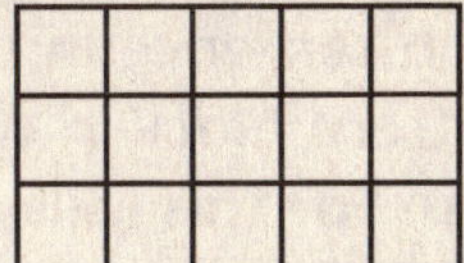

○ **F** 32

○ **G** 48

○ **H** 60

○ **J** 72

Directions: Use your counters for the next question. Read the question carefully and write your answer on the line provided.

3 **Measure the rectangle below with your counters. How many counters would be needed to cover $\frac{1}{2}$ of the rectangle? Write your answer on the line below.**

Answer ____________ **counters**

MILE 26: SPATIAL SENSE

Spatial sense is the ability to recognize how shapes relate to or fit into one another. You may have to make figures with your pattern blocks. You may also be asked how many pattern blocks are needed to cover another figure or part of a figure.

Directions: Each sentence in the left column describes a pattern block. Find the letter of the correct pattern block in the right column and write it in front of the description in the left column. For this Mile, use the pattern blocks cut from page 7.

_____ 1. This pattern block is $\frac{1}{2}$ the size of pattern block A.

_____ 2. This pattern block is $\frac{1}{2}$ the size of pattern block B.

_____ 3. This pattern block is $\frac{1}{3}$ the size of pattern block E.

_____ 4. This pattern block is $\frac{1}{3}$ the size of pattern block A.

_____ 5. This pattern block is $\frac{1}{6}$ the size of pattern block A.

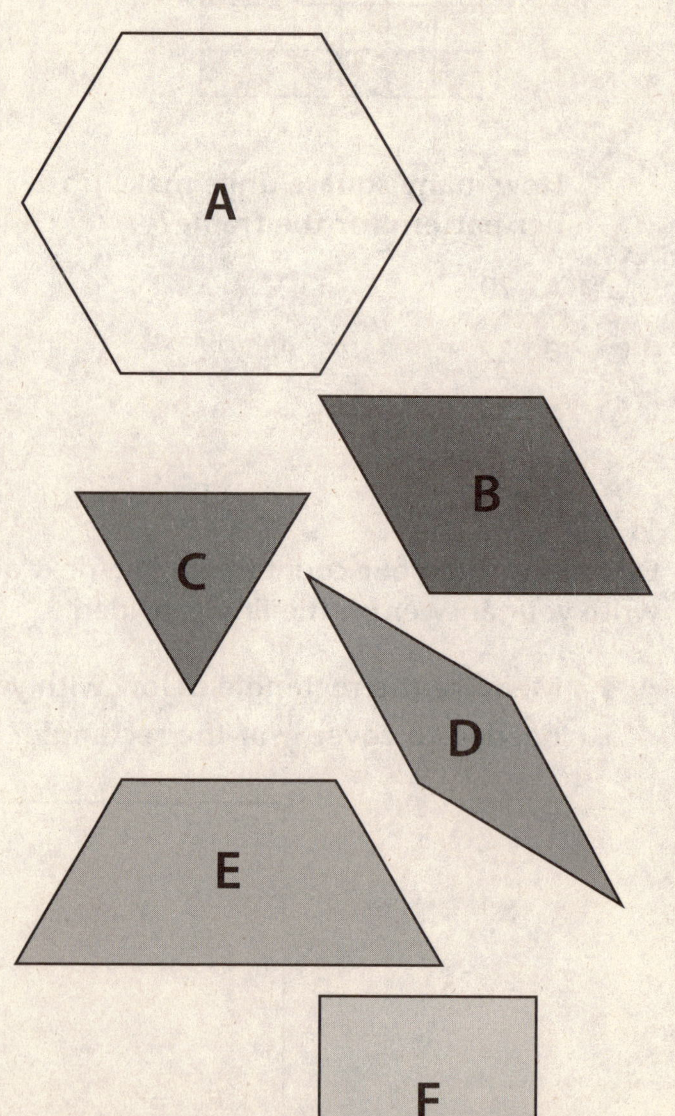

Directions: Read each question carefully. Use your pattern blocks to answer these questions. Fill in the bubble next to your answer choice for each question.

1 **Place pattern block E and pattern block C together so that two sides are touching. What shape can you get?**

○ **A** rectangle

○ **B** trapezoid

○ **C** rhombus

○ **D** parallelogram

Use the figure below to answer questions 2 and 3.

2 **Which of the following combinations of pattern blocks can be used to make the figure?**

○ **F** 5 of pattern block C and 2 of pattern block E

○ **G** 1 of pattern block A and 5 of pattern block C

○ **H** 1 of pattern block A and 2 of pattern block E

○ **J** 4 of pattern block B and 5 of pattern block C

3 **How many of pattern block C would you need to cover $\frac{1}{2}$ of the figure?**

○ **A** 4

○ **B** 6

○ **C** 9

○ **D** 12

Mile 27: Grids

Grids are systems of vertical and horizontal lines. Points on a grid are located using **ordered pairs** of numbers. Starting at 0 on the grid, first go *right* to the first number in the ordered pair. Then, go *up* to the second number in the ordered pair.

There is a point on the grid at right. Look at the numbers at the bottom of the grid to find the first number of the ordered pair. It is 2. Look at the numbers at the left of the grid to find the second number of the ordered pair. It is 3. The ordered pair is (2, 3).

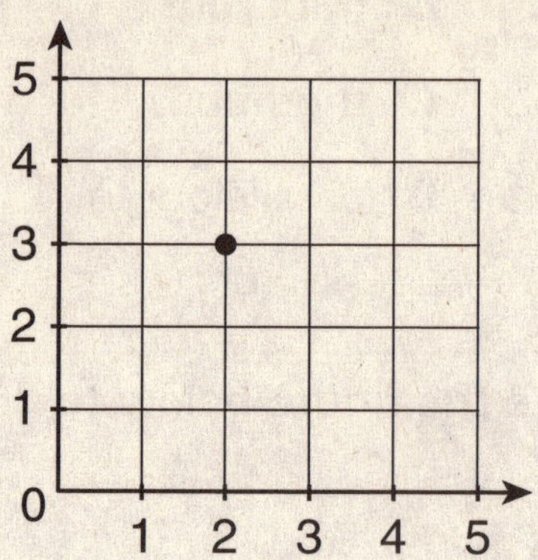

Directions: Each point on the grid below is labeled with a letter. The boxes below the grid show the ordered pairs for different points. Find each point and then fill in the letter that goes with it in the box. The solution spells out the answer to the riddle below.

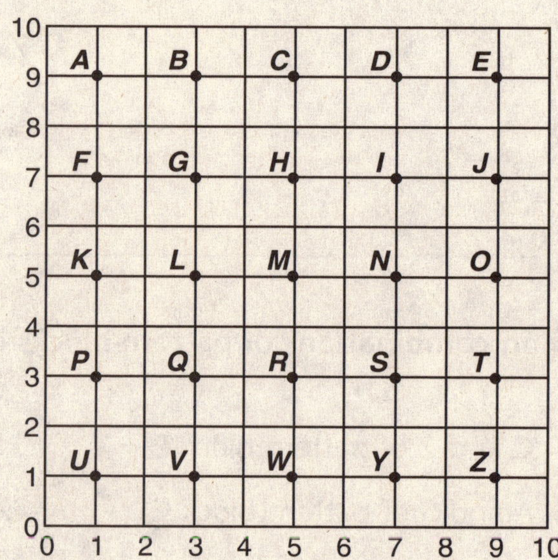

Who is bigger, Mrs. Bigger or Mrs. Bigger's baby?

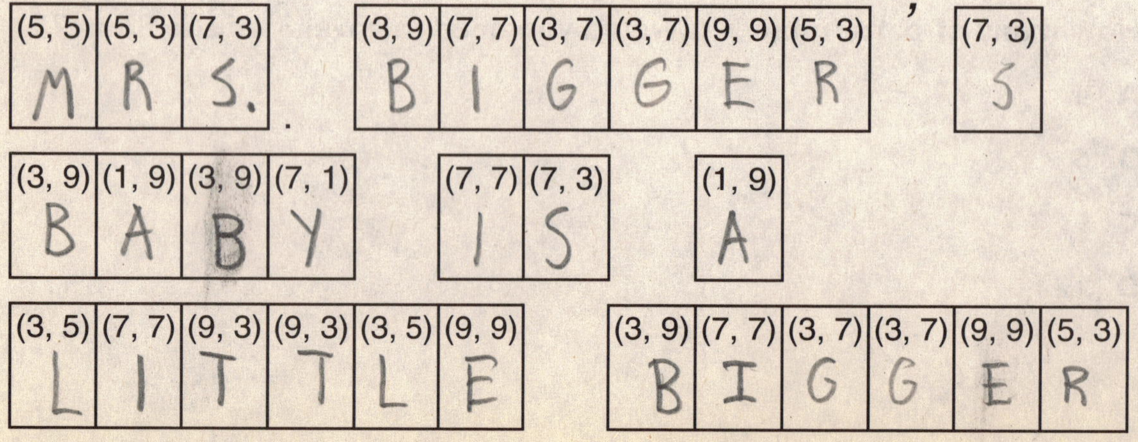

Directions: Read each question carefully. Fill in the bubble next to your answer choice for each question.

1 **Which set of ordered pairs describes the endpoints of the line below?**

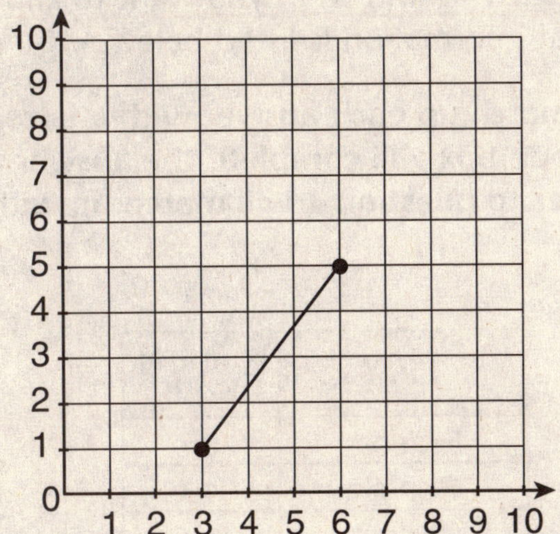

● **A** (3, 1), (6, 5)

○ **B** (1, 3), (6, 5)

○ **C** (1, 3), (5, 6)

○ **D** (3, 1), (3, 5)

2 **Tanya and Janie have been given a map to get around town during their vacation. If they keep walking in the same pattern, what point will they get to?**

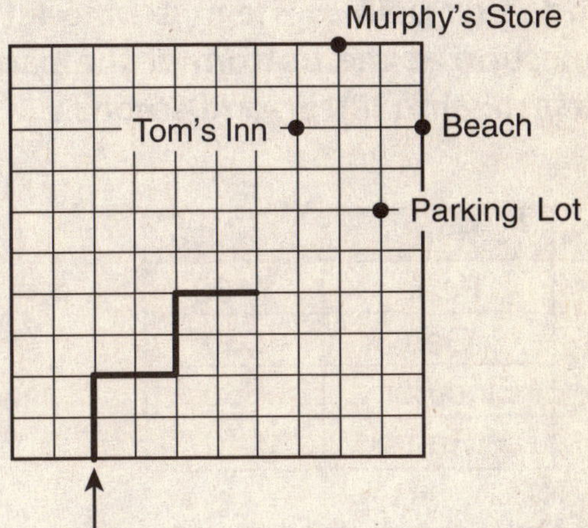

○ **F** Murphy's Store

○ **G** Tom's Inn

○ **H** Beach

● **J** Parking Lot

Directions: Read the question carefully and write your answers on the lines provided.

3 **There is a square on the grid below. Write the ordered pairs of each corner of the square on the lines below.**

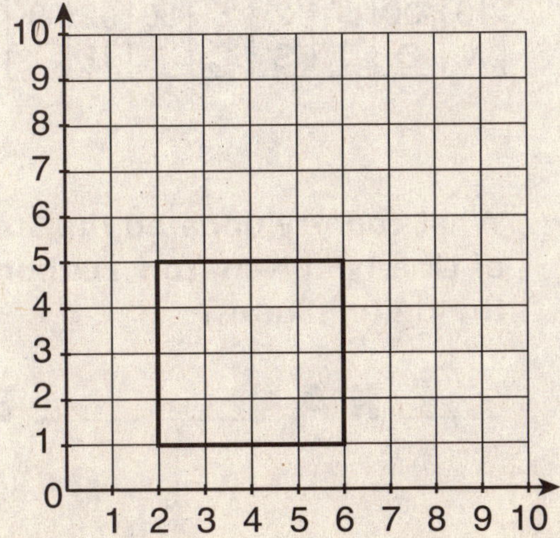

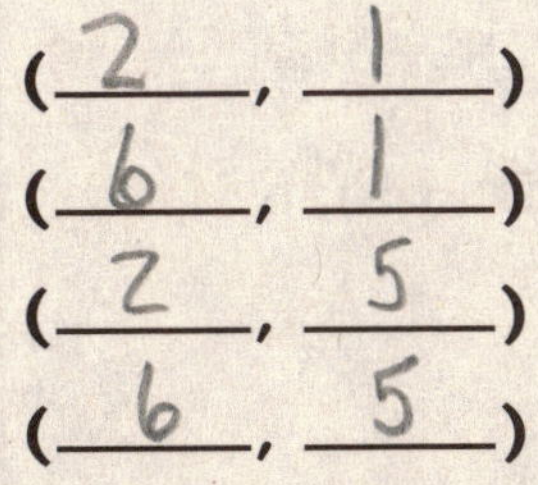

A common mistake in finding the point of an ordered pair is reversing the numbers. Make sure that you go *right* for the first number and *up* for the second number.

MILE 28: TABLES

A **table** organizes information in a way that makes it easy to find what you want to know. The headings at the top of each column tell you what information is listed below.

Directions: Answer the following questions. Then match up each answer with a letter in the code box below. The matching letters in the code box will complete the answer to the question at the bottom of the page. The answer to question 1 will match up with the first missing letter, and so on.

1.

Expenses	Dollars
Food	$100
Gas	$30
Electricity	$60
Fun	$40

How much money was spent on gas each week?

$__________

2.

Family Member	Height (cm)
Mom	172
Dad	180
Mei Su	164
Lam	154

How much taller is Dad than Mei Su?

__________ cm

3.

Item	Number Sold
Erasers	25
Pencils	40
Markers	12
Notebooks	20
Book Covers	18
Folders	14

How many more erasers than notebooks were sold?

____

Code Box

16	100	60	2	5	30	92	3
N	O	T	S	C	F	U	I

► **What country donated the Statue of Liberty in New York Harbor to the United States?**

____ R A ____ ____ E

Directions: The table below shows how much money Keith saved each month. Use the information in the table to answer the questions below. Read each question carefully. Fill in the bubble next to your answer choice for each question.

Month	Money Saved
Jan.	$22.50
Feb.	$40.15
March	$33.22
April	$26.18
May	$40.15
June	$15.17

1 In which month did Keith save the least?

- ○ **A** January
- ○ **B** March
- ○ **C** May
- ○ **D** June

2 Which of the following statements is true about Keith's savings record?

- ○ **F** Keith saved more money in January than in February.
- ○ **G** Keith saved less money in April than in March.
- ○ **H** Keith saved more money in June than in January.
- ○ **J** Keith didn't save any money in May.

3 During which two months did Keith save the same amount of money?

- ○ **A** February and May
- ○ **B** March and April
- ○ **C** January and May
- ○ **D** May and June

4 About how much did Keith save during January and February?

- ○ **F** $50
- ○ **G** $60
- ○ **H** $70
- ○ **J** $75

MILE 29: BAR GRAPHS

Bar graphs use bars of different heights or lengths to compare information. Each bar graph has numbers along one side. You can find how much each bar represents by looking at the number that it reaches.

Directions: Count the items below the graph. Show how many there are of each item by filling out the bar graph. The first one has been done for you. Choose a different color for each bar. Title and label the axes on the graph.

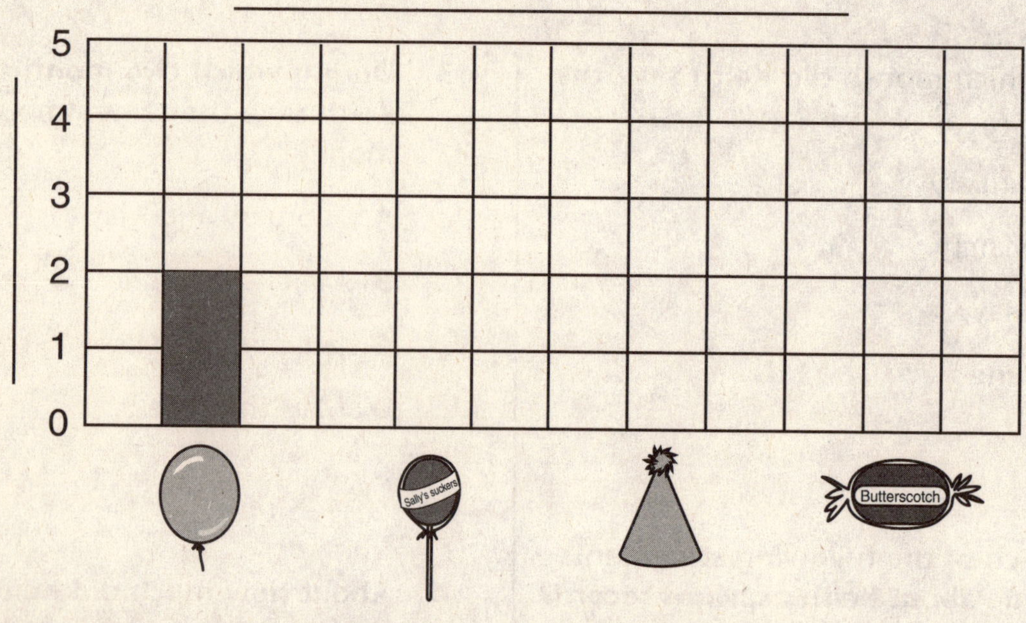

Directions: The bar graph shows the fruit that was purchased at Caldwell Cafeteria last Thursday. Read each question carefully. Look at the graph and fill in the bubble next to your answer choice for each question.

1 Caldwell Cafeteria had a supply of 40 pears at the beginning of the day. How many pears were not purchased by the end of the day?

○ **A** 5

○ **B** 10

○ **C** 20

○ **D** 35

2 What was the total number of pieces of fruit sold at Caldwell Cafeteria?

○ **F** 40

○ **G** 80

○ **H** 115

○ **J** 130

Directions: Read the question carefully and use the information from Kendra's journal to make a bar graph.

3 Kendra kept track of the number of seeds of each vegetable she planted in her garden. Look at the page from her planting journal.

Squash .... 25 seeds
Cucumbers .... 50 seeds
Eggplant .... 20 seeds
Tomatoes ...... 15 seeds
Corn .... 50 seeds

Use the record that Kendra kept to draw a bar graph in the grid. Title and label the bar graph.

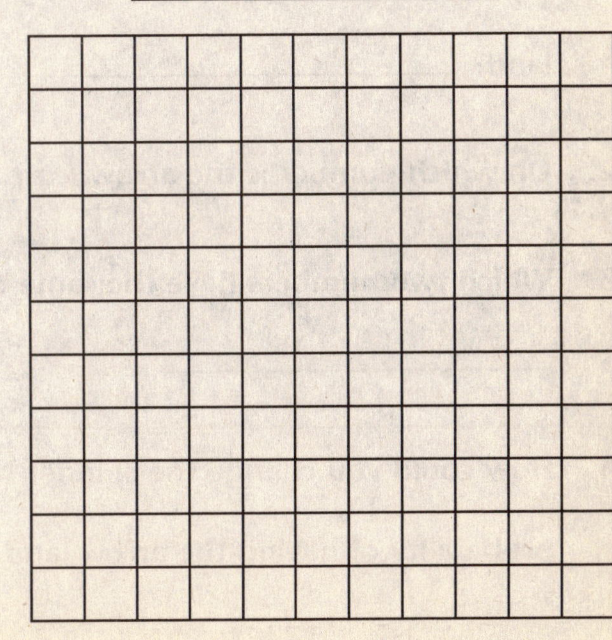

MILE 30: PROBABILITY

Probability tells you how likely it is that an event will happen. Some events are more likely to happen than other events. Say that there are 6 red marbles, 3 blue marbles, and 1 green marble in a bag. If you picked one marble without looking, you are *more likely* to get a red marble than a blue or green marble. That's because $\frac{6}{10}$ of the marbles in the bag are red.

Directions: Sam and Lisa are playing a board game. The different numbers on the spinner will tell Sam and Lisa how many spaces to move their game pieces. The spinner is divided into 8 equal sections. Look at the spinner carefully. Then read and answer the questions.

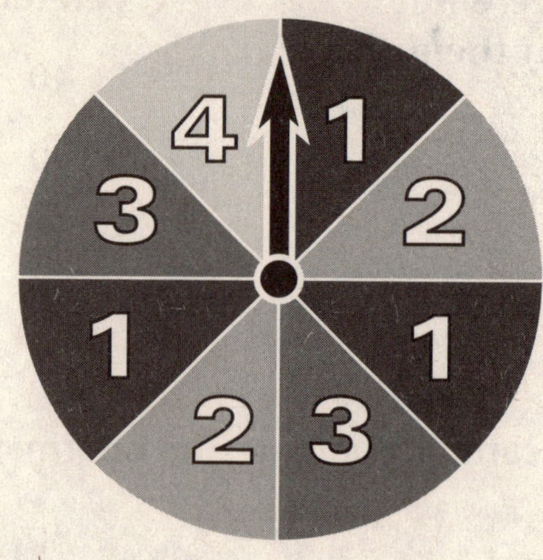

1. **If Sam or Lisa takes a turn spinning, on which number will the arrow *most likely* land?** 1

2. **On which number is the arrow *least likely* to land?** 4

3. **Which two numbers have the same chances of the arrow landing on them?** 2 and 3

4. **How could you change the spinner so that all four numbers have the same probability of having the arrow land on them?** Each number has two spots.

Directions: Read each question carefully. Fill in the bubble next to your answer choice for each question.

1 A cube has 6 sides. There are 3 blue sides, 2 red sides, and 1 yellow side. If the cube is rolled, what is the probability that it will land red side up?

- ○ **A** $\frac{1}{6}$
- ● **B** $\frac{2}{6}$
- ○ **C** $\frac{3}{6}$
- ○ **D** $\frac{4}{6}$

2 Monica has 8 gym balls in a mesh bag. There are 3 yellow balls, 2 green balls, 2 red balls, and 1 blue ball. If she takes out a ball randomly, which color ball has a 3 out of 8 probability of being chosen?

- ● **F** yellow
- ○ **G** green
- ○ **H** red
- ○ **J** blue

Directions: Read the question below and write your answer on the line provided.

3 Jose is playing a game with Derek. If the arrow points to a piece of fruit, Jose will win the game.

Spinner 1

Spinner 2

Spinner 3

Which spinner is most likely to have the arrow point to a piece of fruit?

Answer Spinner 3

Explain why this spinner will give Jose the best probability of landing on a piece of fruit.

This spinner will give Jose the best probability of landing on a piece of fruit because the biggest spot on the spinner is a fruit.

The total number of possible events will be the bottom number in a probability. The top number will be the number of ways a certain event can happen.

MILE 31: NUMBER PATTERNS

A **number pattern** is a series of numbers that have a common relationship. For example, each number in the pattern could be 2 more than the number before it: 2, 4, 6, 8, 10, 12.

You may also notice that this is a list of the multiples of 2: 2×1, 2×2, 2×3, and so on. Number patterns can use any operation to connect the numbers in the series, as long as it is consistent.

Directions: Create number patterns using the scores of the Women's National Basketball Association's New York Liberty basketball team and the National Football League's New York Giants football team.

1. **Imagine that the Liberty made 8 baskets worth 3 points each to start a basketball game. How would their score look after each basket? Finish the number pattern below to find out.**

   0, 3, 6, 9, _____, _____, _____, _____, _____

2. **Now imagine that the Liberty had 31 points at halftime. If they scored 6 baskets worth 2 points each to start the half, how would their score look after each basket? Finish the number pattern below to find out.**

   31, 33, 35, _____, _____, _____, _____

3. **Imagine that the Giants scored 6 touchdowns in a game. For each touchdown, they got the extra point and scored 7 points. How would their score look after each touchdown? Finish the number pattern below to find out.**

   0, 7, 14, _____, _____, _____, _____

4. **Now imagine that the Giants were playing another game. They scored a 3-point field goal and then a 7-point touchdown to start the game. They continued to score in this pattern for the whole game. How would the score look after each field goal and touchdown? Finish the number pattern below until you run out of spaces to write.**

   0, 3, 10, 13, 20, _____, _____, _____, _____

   The Giants were playing the Dallas Cowboys in this game. If the Cowboys scored 37 points in the game, which team won?

   ____

Directions: Read each question carefully. Fill in the bubble next to your answer choice for each question.

1 What is the next number in the pattern below?

15, 20, 25, 30, 35, 40, ______

○ A 35

○ B 40

○ C 45

○ D 50

2 What is the next number in the pattern below?

$9\frac{1}{2}$, 9, $8\frac{1}{2}$, 8, $7\frac{1}{2}$, ______

○ F 6

○ G $6\frac{1}{2}$

○ H 7

○ J $7\frac{1}{2}$

3 What number is missing from this pattern?

50, 42, 35, ______, 24

○ A 28

○ B 29

○ C 30

○ D 31

Directions: Read each question carefully and write your answer on the line provided.

4 The numbers below form a pattern. Describe the pattern.

95, 102, 109, 116, 123

____

5 Describe the pattern in this number pattern.

57, 51, 46, 42, 39

____

Skip-counting can help you answer questions about number patterns. When you skip-count, you count by 2s, 3s, 4s, 5s, or whatever other number you choose.

Mile 32: Geometric Patterns

A **geometric pattern** is a set of shapes that are related to one another. The pattern might continue in a line, like Example 1, or it may continue out in two dimensions, like Example 2. Once you recognize a pattern, you can guess what shape will come next.

Example 1: **Example 2:**

Directions: Complete the patterns shown below. For each pattern, draw the missing figure.

1. ____________

2. ________

3. 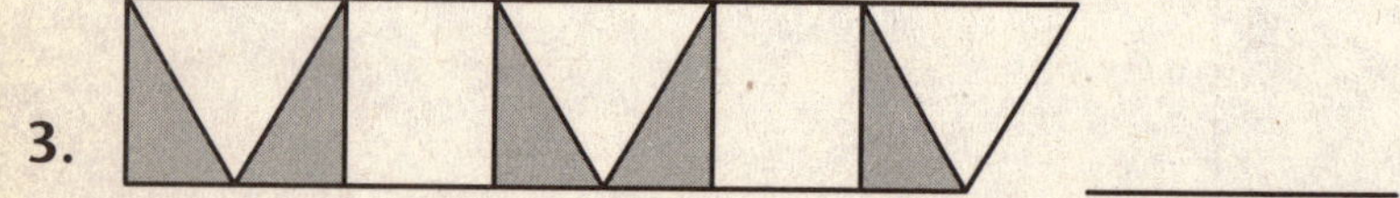 ________

4. ____________

5. 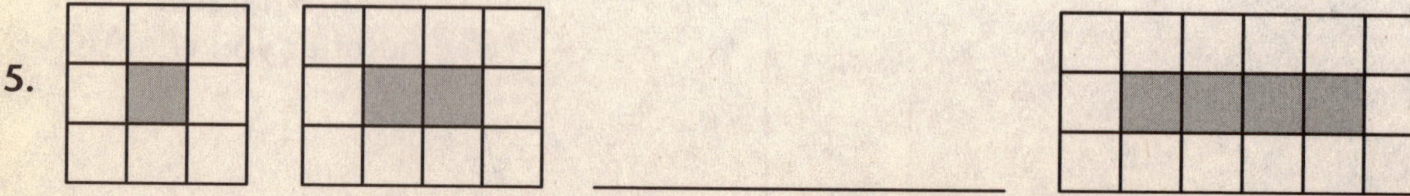

Directions: Read each question carefully. Fill in the bubble next to your answer choice for each question.

1 Look at the pattern below. Which of the following would come next in the pattern?

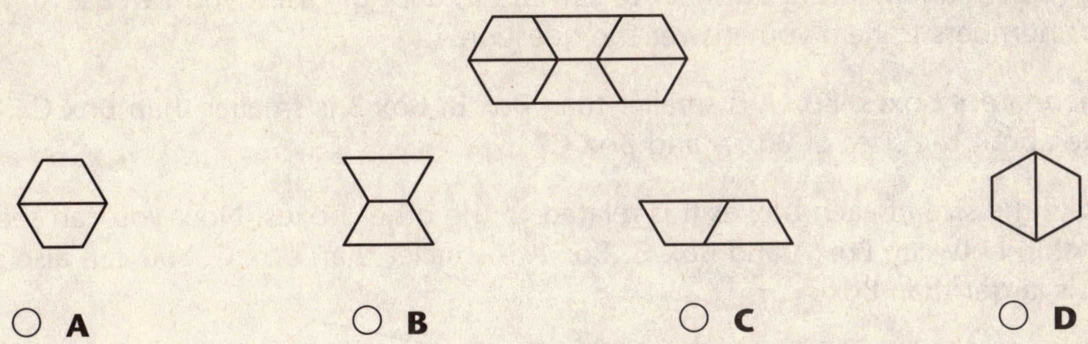

○ A ○ B ○ C ○ D

2 Look at the pattern below. Which of the following would come next in the pattern?

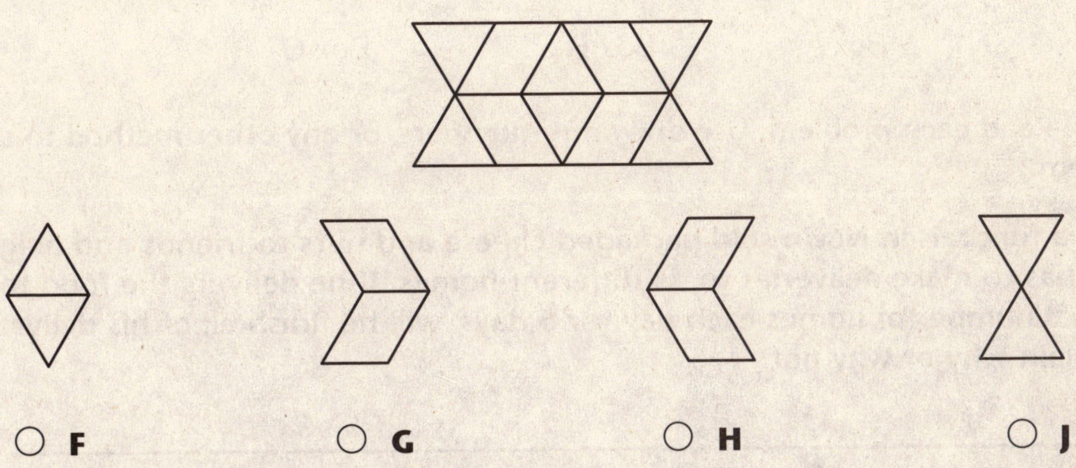

○ F ○ G ○ H ○ J

Directions: Read the question carefully and write your answer in the space provided.

3 Look at the pattern below. Draw the shape that will come next in the pattern.

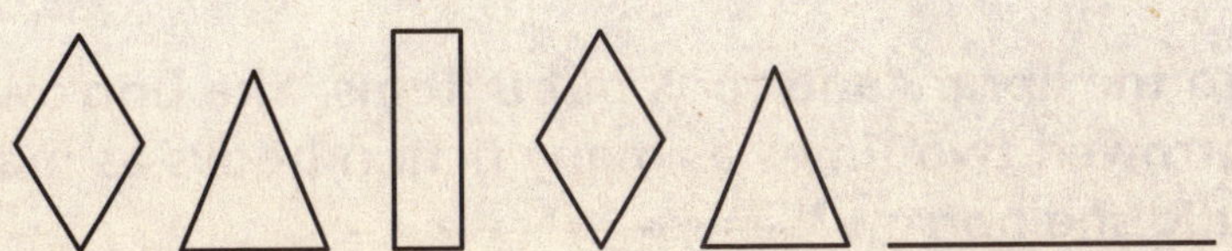

Mile 33: Reasoning

Some questions require you to use **reasoning** to find the answers. Reasoning is thinking about the problem and figuring out how to solve it. In a lot of cases, you can use drawings or numbers to help you answer the question.

Try this: There are 3 boxes. Box A is smaller than Box B. Box B is smaller than Box C. What is true about the sizes of Box A and Box C?

You can draw the size of each box as it is related to the other boxes. Now you can see the relationship between Box A and Box C. Box A is smaller than Box C. You can also say that Box C is larger than Box A.

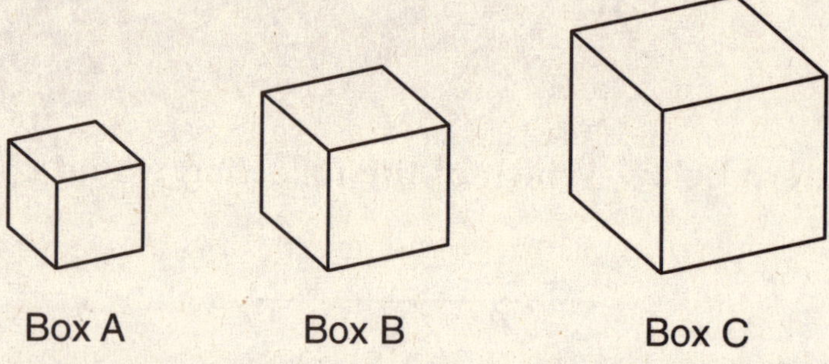

Directions: Read each problem. Use drawings, numbers, or any other method to solve the problem.

1. **For a fundraiser, Nevin sold packaged cheese and nuts to friends and neighbors. He has to make deliveries to 44 different homes. If he delivers the food to an equal number of homes each day for 6 days, will he finish all of his deliveries? Explain why or why not.**

   ____

2. **If you add together 3 odd numbers, will the answer be even or odd? Explain how you know.**

   ____

3. **There were 4 teams in a relay race. Team C finished between Teams A and B. Team D finished before Team A. Team B did not finish first. Who won the race?**

   ____

4. **Lauren went to the library and took out 8 items. She borrowed 2 nonfiction books. She borrowed two times as many fiction books as magazines. How many fiction books did she borrow?**

   ____

Directions: Read each question carefully. Fill in the bubble next to your answer choice for each question.

1 **Mr. Stringer is a tailor. The table below shows the items he can make and the yards of fabric needed to make each item.**

Item	Yards of Fabric
Blanket	12
Suit	9
Dress	4
Shirt	1

If Mr. Stringer has 14 yards of fabric, what items can he make and have no fabric left over?

○ **A** a blanket, a dress, and a shirt

○ **B** a suit and a blanket

○ **C** a dress and a suit

○ **D** a suit, a dress, and a shirt

2 **There are 9 bagels to share between 4 people. If each person eats the same number of bagels and none are cut into pieces, how many bagels will not be eaten?**

○ **F** 1

○ **G** 4

○ **H** 6

○ **J** 8

3 **There are 17 children going to the county spelling bee. Only 4 children can fit into a car. What is the minimum number of cars needed to transport the children to the county spelling bee?**

○ **A** 4

○ **B** 5

○ **C** 8

○ **D** 17

Directions: Read the question carefully and write your answer on the lines provided.

4 **Kelly has 2 hours before soccer practice. She has 30 minutes of reading, 45 minutes of clarinet practice, 45 minutes of math homework, and 60 minutes of science homework. What work can Kelly complete before soccer practice? Explain your answer.**

____

____

____

When solving a reasoning problem, write down the given information and think about how you can use it to solve the problem. If it is a multiple-choice question, the answer choices may help you to find the answer.

Answer Key for Miles

Mile 3

1. M
2. A
3. D
4. I
5. S
6. O
7. N

Madison is the name of the fourth president and the capital of Wisconsin.

---

1. C
2. G
3. D
4. F
5. 7
6. 329,005

Mile 4

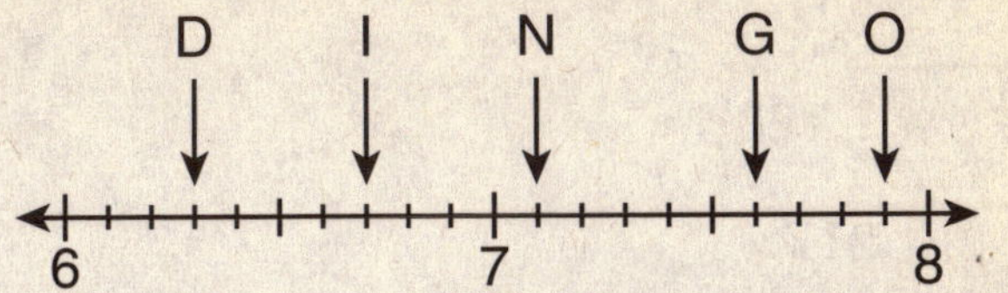

Dingo is the name of a wild dog in Australia.

---

1. C
2. H
3. A
4. G
5. 897, 898, or 899
6. 10.3

Mile 5

1. 541
2. 1,128
3. 217
4. 1,044
5. 904
6. 490
7. 518
8. 981

G	A	L	A	X	Y
541	217	1,044	217	490	981

---

1. C
2. J
3. C
4. G
5. B
6. F

Mile 6

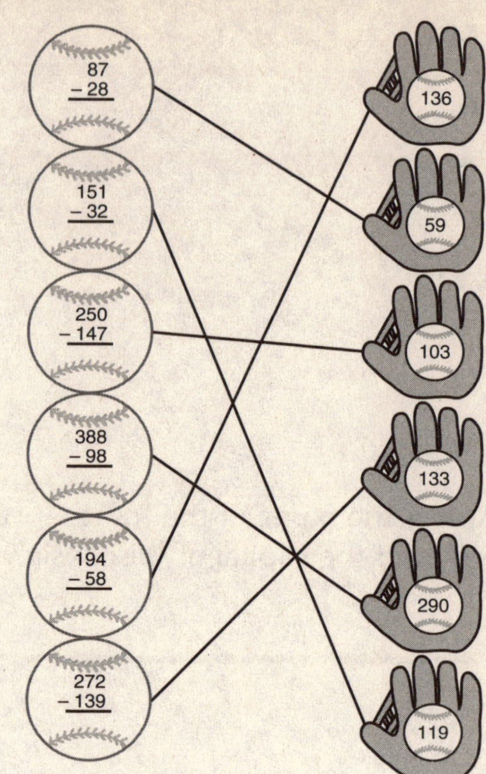

87 − 28 = 59

151 − 32 = 119

250 − 147 = 103

388 − 98 = 290

194 − 58 = 136

272 − 139 = 133

---

1. C
2. G
3. D
4. H
5. A
6. H

Mile 7

1. I = 162
2. S = 460
3. A = 111
4. A = 525
5. C = 308
6. N = 414
7. E = 112
8. W = 384
9. T = 415
10. O = 216
11. N = 64

The name of the famous mathematician is **Isaac Newton.**

---

1. D
2. G
3. C
4. F
5. B
6. H

Mile 8

1. 9
2. 3
3. 6 r6
4. 2
5. 8
6. 12
7. 12 r2
8. 14 r1
9. 43
10. 31 r3
11. 31
12. 10 r7

Why was the archaeologist upset?

His job was in ruins.

---

1. C
2. G
3. A
4. J
5. D
6. G
7. C

Mile 9

	4	5	6	7
21	P	O	M	(A)
45	W	(N)	I	D
18	X	S	(T)	F
32	(A)	K	U	E
25	T	(R)	A	Y
49	Q	U	L	(C)
66	H	N	(T)	V
16	(I)	D	O	B
50	J	(C)	Z	I
54	S	E	(A)	R

The name of one of the seven continents is **Antarctica.**

---

1. B
2. J
3. C
4. J
5. A

Mile 10

$9 \times 5 = 45$

$18 + 9 = 27$

$45 \div 9 = 5$

$27 - 9 = 18$

1. $40 \div 5 = 8$ or $40 \div 8 = 5$
2. $13 + 41 = 54$ or $41 + 13 = 54$
3. $34 - 11 = 23$ or $34 - 23 = 11$
4. $84 \div 12 = 7$ or $84 \div 7 = 12$
5. $12 + 7 = 19$ or $7 + 12 = 19$
6. $2 \times 18 = 36$ or $18 \times 2 = 36$

---

1. C
2. J
3. B
4. $6 \times 3 = 18$

   $18 \div 3 = 6$
5. $8 + 5 = 13$

   $13 - 5 = 8$

Mile 11

1. 48 dumplings
2. 8 servings
3. 14 tacos
4. 8 slices
5. 9 dumplings

---

1. C
2. F
3. B
4. J
5. 8 sets
6. $168

Mile 12

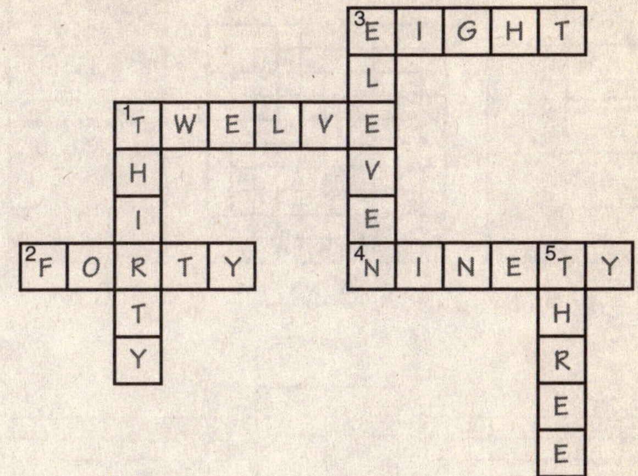

---

1. A
2. H
3. C
4. J
5. 18 red cards
6. 53 inches

Mile 13

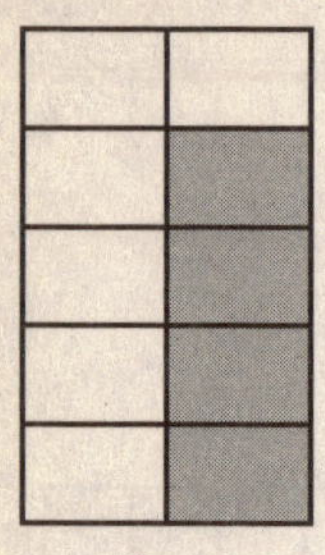
0.4

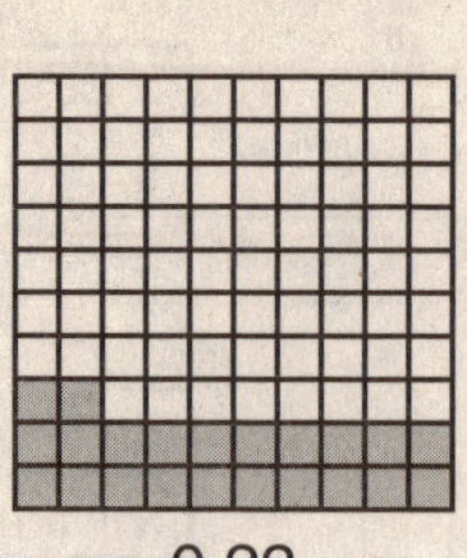
0.22

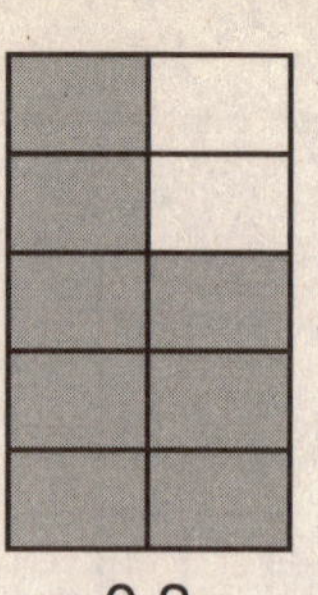
0.8

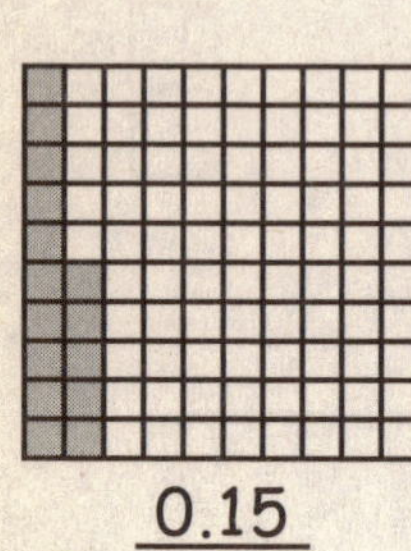
0.15

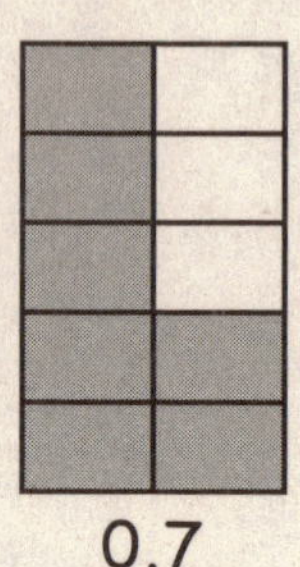
0.7

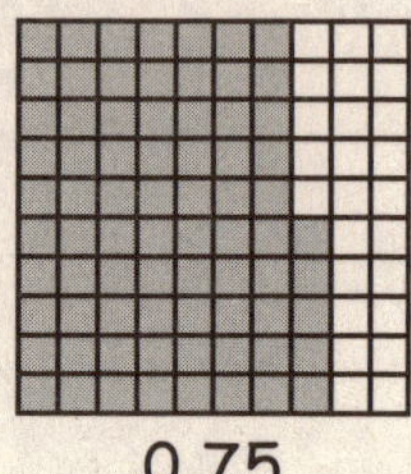
0.75

1. C
2. J
3. B
4. F
5. 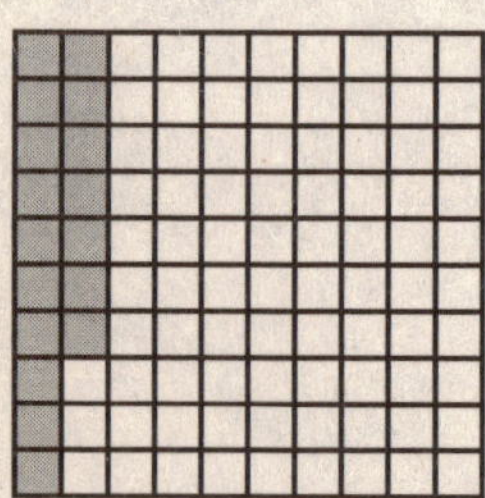

Mile 14

1. 45%
2. 50%
3. 30%
4. 10%
5. 80%
6. 37%

---

1. A
2. H
3. D
4. J
5. 53%, 54%, 60%, 62%, 67%

Mile 15

1. 6
2. $\frac{2}{6}$
3. $\frac{3}{6}$
4. $\frac{1}{6}$

---

1. B
2. J
3. C
4. $\frac{4}{8}$
5. $\frac{6}{10}$

Mile 16

1. $\frac{1}{5} = $ A
2. $\frac{1}{4} = $ Penny
3. $\frac{4}{10} = $ Saved
4. $\frac{1}{2} = $ Is
5. $\frac{2}{3} = $ A
6. $\frac{5}{6} = $ Penny
7. $\frac{7}{8} = $ Earned

Benjamin Franklin said, **"A penny saved is a penny earned."**

---

1. D
2. J
3. B
4. $\frac{5}{8}, \frac{3}{8}, \frac{2}{8}, \frac{1}{8}$
5. $\frac{3}{10}, \frac{3}{8}, \frac{3}{5}, \frac{3}{4}$
6. $\frac{1}{5} < \frac{1}{4}$

   $\frac{3}{6} = \frac{1}{2}$

   $\frac{4}{6} > \frac{4}{8}$

Mile 17

1. 5
2. 1
3. 7
4. 6
5. 8
6. 2
7. 4
8. 3

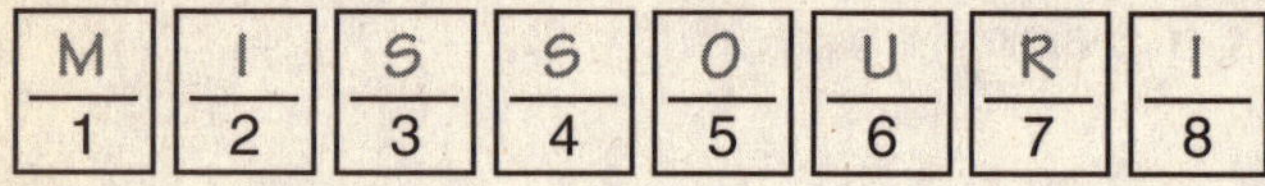

1. B
2. F
3. C
4. Lois had the most plants left. She had 4 more plants.
5. 3 female guppies
6. 2 black socks

Mile 18

1. 50
2. 90
3. 240
4. 170
5. 500
6. $\frac{5}{10}$
7. $\frac{8}{10}$
8. $\frac{3}{10}$
9. 190 + 80 = 270
10. 100 × 10 = 1,000
11. $10\overline{)60} = 6$

1. B
2. H
3. A
4. Michael
5. 24 ounces

Mile 19

Length

1 foot = 12 inches

1 yard = 3 feet

1 meter = 100 centimeters

Weight

1 pound = 16 ounces

1 kilogram = 1,000 grams

Liquid Volume

1 pint = 2 cups

1 quart = 2 pints

1 gallon = 4 quarts

1 liter = 1,000 milliliters

1. T
2. F
3. F
4. F
5. F

---

1. C
2. G
3. A
4. H
5. B
6. J

Mile 20

1. 5 cm, 2 in.
2. 7 cm, $2\frac{3}{4}$ in.
3. $7\frac{1}{2}$ cm, 3 in.
4. $2\frac{1}{2}$ cm, 1 in.
5. 14 cm, $5\frac{1}{2}$ in.

---

1. C
2. H
3.

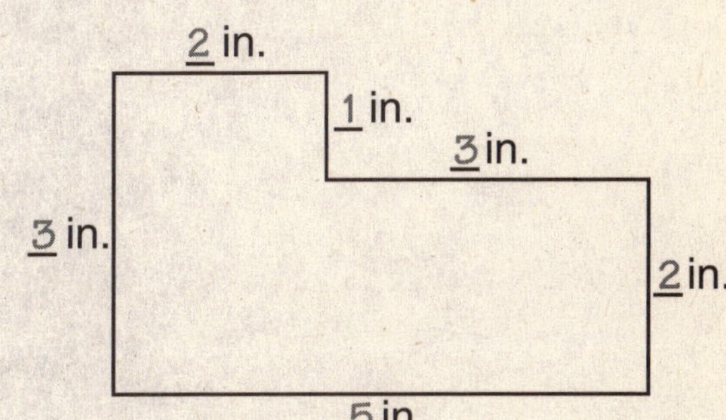

Mile 21

You should have crossed out the first row (B), the fourth row (C), the sixth row (K), the ninth row (N), and the eleventh row (D).

The answer to the riddle is **lawsuit.**

---

1. A
2. H
3. C
4. 5:45 P.M.
5. 2:56 P.M.

Mile 22

1. Square
2. Parallelogram
3. Triangle
4. Rhombus
5. Trapezoid
6. Circle
7. Rectangle
8. Hexagon

Circle = 1

Triangle = 2, 4, 7, 8

Trapezoid = 3

Parallelogram = 6

Rectangle = 5, 9, 10, 12, 13

Square = 11

---

1. D
2. H
3. D
4. Trapezoid

Mile 23

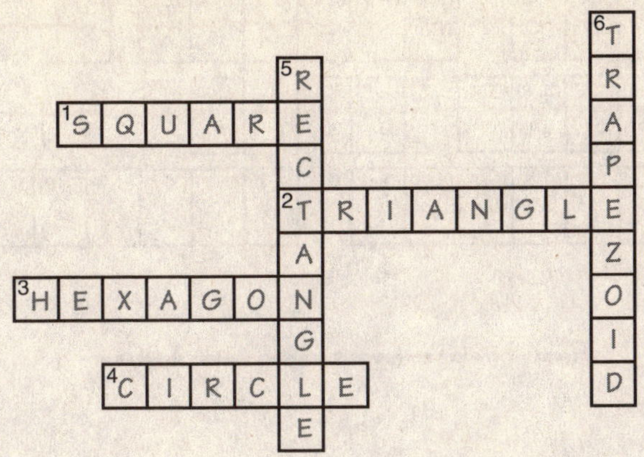

1. C
2. F
3. B

Mile 24

1. 26 centimeters
2. 20 inches
3. 32 centimeters
4. 41 centimeters
5. 12 inches

C	R	A	N	E
12 in.	41 cm	20 in.	26 cm	32 cm

1. D
2. H
3. 36 centimeters

Mile 25

1. Any figure that has an area of 9 square units is correct. Count the units in your drawing to make sure there are 9.
2. Any figure that has an area of 12 square units is correct. Count the units in your drawing to make sure there are 12.
3. Any figure that has an area of 10 square units is correct. Count to make sure that you have drawn 10 units.

1. A
2. H
3. 3 counters

Mile 26

1. E
2. C
3. C
4. B
5. C

---

1. D
2. H
3. B

Mile 27

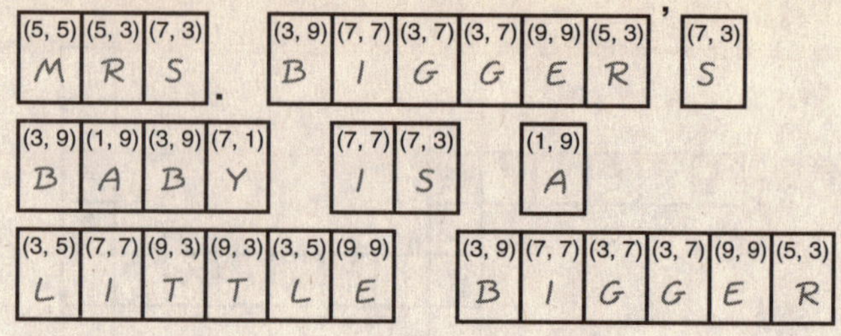

---

1. A
2. H
3. (2, 1), (2, 5), (6, 1), (6, 5)

Mile 28

1. $30

2. 16 cm

3. 5

France donated the Statue of Liberty.

---

1. D

2. G

3. A

4. G

Mile 29

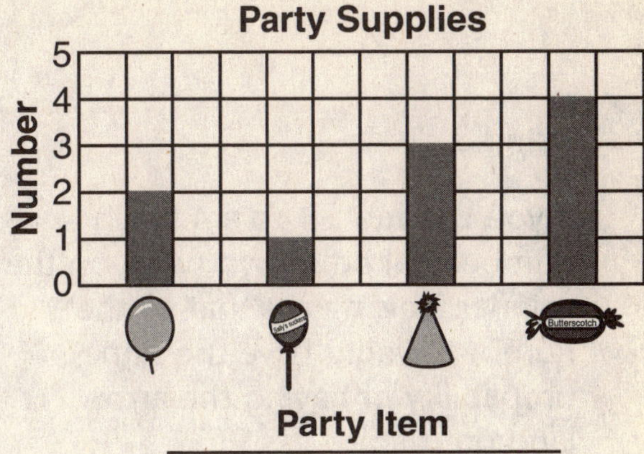

---

1. A

2. H

3. Kendra's Garden

Number of Seeds
55
50
45
40
35
30
25
20
15
10
5
0
Squash
Cucumbers
Eggplant
Tomatoes
Corn
Vegetables

Mile 30

1. 1
2. 4
3. 2 and 3
4. If you change a 1 to a 4, each number will be shown twice on the spinner. This means that all the numbers would have the same probability of having the arrow land on them.

---

1. B
2. F
3. Spinner 3

   Spinner 3 gives Jose the best probability of landing on a piece of fruit because the fruit sections cover more than $\frac{1}{2}$ of the spinner. The fruit sections on the other two spinners are $\frac{1}{2}$ or less.

Mile 31

1. 0, 3, 6, 9, 12, 15, 18, 21, 24
2. 31, 33, 35, 37, 39, 41, 43
3. 0, 7, 14, 21, 28, 35, 42
4. 0, 3, 10, 13, 20, 23, 30, 33, 40

   The Giants won the game, 40–37.

---

1. C
2. H
3. B
4. Add 7 to each number.
5. Subtract 6, 5, 4, 3.

Mile 32

1. 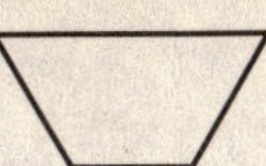

2. 

3. 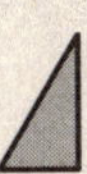

4. 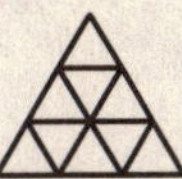

5. 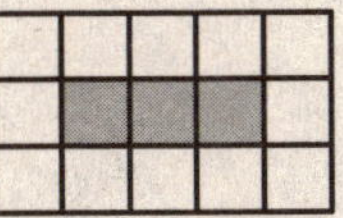

---

1. B
2. H
3. 

Mile 33

1. Nevin cannot deliver food to an equal amount of homes each day. If he delivers food to 8 homes per day for 6 days, he will finish all of his deliveries ($8 \times 6 = 48$). But if he delivers food to 7 homes per day for 6 days, he will not finish all of his deliveries ($7 \times 6 = 42$).
2. The answer will be odd. $3 + 3 + 3 = 9$. $7 + 7 + 7 = 21$. When three odd numbers are added together, the sum is odd.
3. Team D
4. Lauren borrowed 4 fiction books.

---

1. D
2. F
3. B
4. Kelly can finish her reading, clarinet practice, and math homework before she leaves for soccer practice. $30 + 45 + 45 = 120$. 120 minutes is 2 hours.

Practice Tests

How to Take the Practice Tests

Try to take each practice test as if it were the actual New York State Grade 4 Math test. Take one session at a time over three days. If you begin a practice test, don't stop until you are finished. You should also time yourself while you take the tests so that you'll know what it feels like to be timed as you take the actual test.

- Take no more than forty minutes to answer the questions in Session 1 of each practice test.
- Take no more than fifty minutes to answer the questions in Session 2 of each practice test.
- Take no more than fifty minutes to answer the questions in Session 3 of each practice test.

You will be allowed to write on the test book while you take the New York State Grade 4 Math test, so feel free to write out your work right next to each question that you answer on the practice tests.

Before you take each practice test, you should cut out the answer sheet that goes with it. The answer sheet for Practice Test 1 is on page 103. The answer sheet for Practice Test 2 is on page 143. Each comes right before the practice test begins.

Use all of the tools that you cut out from page 7 while you take the practice tests. These tools include a ruler that measures inches and centimeters, a set of counters, and a set of pattern blocks.

When you take the practice tests, make the setting as realistic as possible. Find a quiet spot where you will not be disturbed. Do not have any open books on the table. Do not watch television, talk on the phone, or listen to music when you take the practice tests. You will **not** be allowed to use a calculator while you take the actual test, so you should not use one while working on the practice tests.

Remember to use the tips and lessons that you have learned and practiced in this book. They will help you do your best on the practice tests. After you have taken each practice test, have an adult go over it. The answers and explanations to the practice tests begin on page 181. Read the explanations of the correct answers for as many questions as you need to. Pay special attention to the explanations of questions that you answered incorrectly or had trouble answering.

Good luck!

Practice Test 1

BOOK 1

Test-Taking Tips

The suggestions below will help you do your best on the test.

- Take the time to read the directions in the Test Book carefully.
- If you do not understand the directions, ask your parent or teacher to explain them to you.
- You should read each question carefully, think about the answer, and then make your response.
- Get rid of wrong answer choices when you don't know the right answer choice for a question.
- Use only No. 2 pencils for the test.
- You may use tools such as your ruler, square counters, or pattern blocks to help you solve some of the test problems.

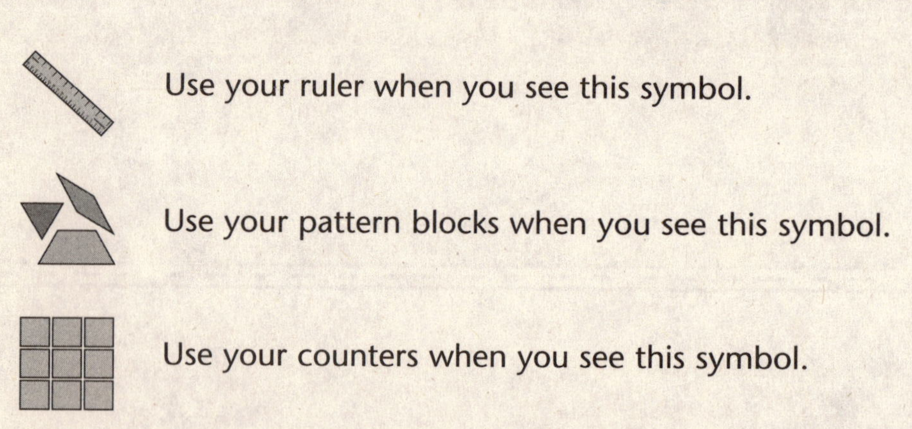

Practice Test 1 Answer Sheet

Completely darken bubbles with a No. 2 pencil. If you make a mistake, be sure to erase mark completely. Erase all stray marks.

1. YOUR NAME: ______________________ Last ______________________ First ______ M.I.
(Print)

SIGNATURE: ______________________ **DATE:** ___ / ___ / ___

HOME ADDRESS: ______________________ Number
(Print)

______________________ City ______________________ State ______ Zip Code

PHONE NO.: ______________________
(Print)

2. YOUR NAME

First 4 letters of last name				FIRST INIT	MID INIT
A–Z	A–Z	A–Z	A–Z	A–Z	A–Z

3. DATE OF BIRTH

Month	Day		Year			
JAN						
FEB						
MAR	0	0	0	0	0	0
APR	1	1	1	1	1	1
MAY	2	2	2	2	2	2
JUN	3	3	3	3	3	3
JUL		4	4	4	4	4
AUG		5	5	5	5	5
SEP		6	6	6	6	6
OCT		7	7	7	7	7
NOV		8	8	8	8	8
DEC		9	9	9	9	9

4. SEX

MALE
FEMALE

The Princeton Review

Practice Test 1

Session 1

Sample A A B C D
Sample B F G H J
Sample C A B C D

1. A B C D
2. F G H J
3. A B C D
4. F G H J
5. A B C D
6. F G H J
7. A B C D
8. F G H J
9. A B C D
10. F G H J
11. A B C D
12. F G H J
13. A B C D
14. F G H J
15. A B C D
16. F G H J
17. A B C D
18. F G H J
19. A B C D
20. F G H J
21. A B C D
22. F G H J
23. A B C D
24. F G H J
25. A B C D
26. F G H J
27. A B C D
28. F G H J
29. A B C D
30. F G H J

Write answers to Questions 31–48 in the space provided.

Session 1

Sample A

$$\begin{array}{r} 866 \\ -\ 38 \\ \hline \end{array}$$

A 808

B 828

C 836

D 838

Sample B

7.5, 8, 8.5, 9, ☐

Which number would be next?

F 8.5

G 9

H 9.5

J 10.5

Go On

Sample C

For this problem, use your ruler.

What is the perimeter of this rectangle?

A 3 inches

B 4 inches

C 5 inches

D 6 inches

1 $\begin{array}{r} 18 \\ \times\ 7 \\ \hline \end{array}$

A 85

B 116

C 126

D 135

2 $\begin{array}{r} 266 \\ +\ 28 \\ \hline \end{array}$

F 284

G 294

H 298

J 304

Go On

3 **Tamara started school at 8:15 A.M. She finished school at 3:00 P.M. How long was she in school?**

A 5 hours, 15 minutes

B 6 hours, 15 minutes

C 6 hours, 45 minutes

D 7 hours, 45 minutes

4 **Which of the following is true about the number 19,340?**

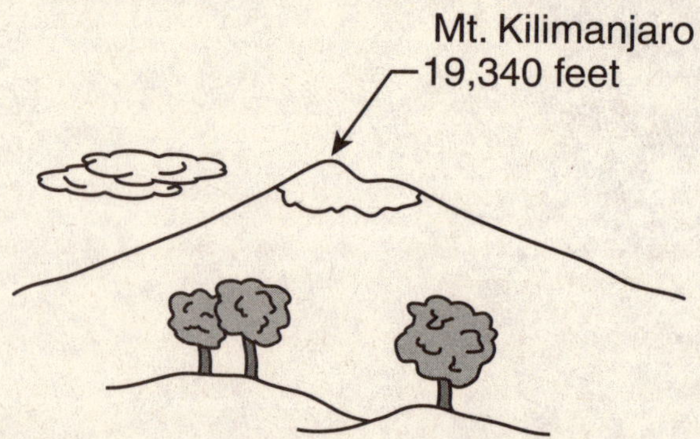

F The 0 is in the tens place.

G The 1 has a value of 1,000.

H The digit in the hundreds place is even.

J The 9 is in the thousands place.

5 **Cara has 6 blue marbles. She has 2 fewer yellow marbles than blue marbles. She has 5 more red marbles than yellow marbles. How many red marbles does she have?**

A 5

B 7

C 9

D 10

6 **About how much does a box of cereal weigh?**

F 12 grams

G 12 ounces

H 12 pounds

J 12 kilograms

7 **Which number below is *greater* than $\frac{1}{4}$?**

A $\frac{1}{3}$

B $\frac{1}{5}$

C $\frac{1}{6}$

D $\frac{1}{8}$

8 **Shawn has \$7 and wants to buy a CD that costs \$10. What percent of the cost of the CD does Shawn have?**

F 0.7%

G 7%

H 70%

J 700%

Go On

9 Marcia counted how many cars crossed the bridge near her house on 4 different mornings between 8:00 A.M. and 9:00 A.M. The graph below shows how many cars crossed the bridge each morning.

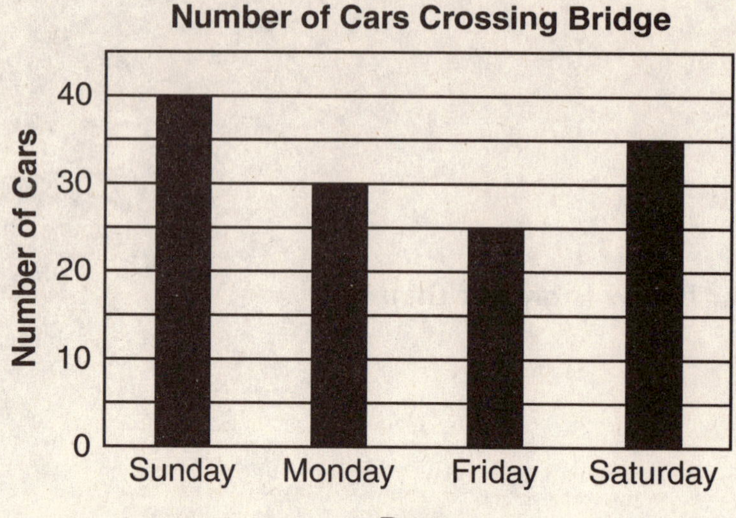

On which day of the week did more than 30 cars but less than 40 cars cross the bridge?

A Sunday

B Monday

C Friday

D Saturday

10 Ms. Hammet had 710 cookies that she wanted to put into 10 boxes. If she tried to divide the cookies evenly, about how many cookies would she put into each box?

F 60

G 70

H 80

J 90

11 Look at the shapes below. Which one has no pairs of parallel lines?

A B C D

12 Susan has 96 stickers that she wants to give equally to 8 people. How many stickers will she give to each person?

F 8

G 12

H 15

J 18

13 Look at the following multiplication problem.

$$\begin{array}{r} 35 \\ \times\ 12 \\ \hline 420 \end{array}$$

Which expression below can be used to find out if the answer is correct?

A 420 + 12

B 420 − 12

C 420 × 12

D 420 ÷ 12

Go On

14 Jeremy has 5 travel cases for his cassette tapes. Each travel case holds 7 tapes. He has 32 tapes. How many *more* tapes can Jeremy buy to completely fill his last travel case?

F 3

G 4

H 5

J 6

15 What fraction of this rectangle is shaded?

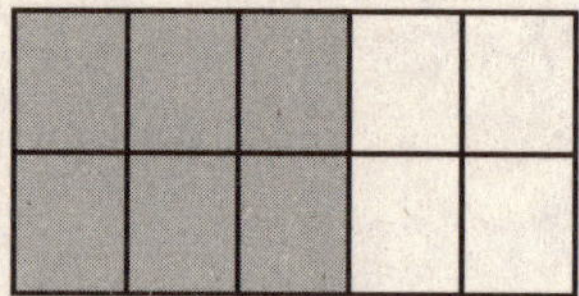

A $\frac{4}{10}$

B $\frac{5}{10}$

C $\frac{6}{10}$

D $\frac{7}{10}$

16 **Which number belongs in the box, according to the pattern?**

42, 36, 30, 24, ☐

F 16

G 18

H 20

J 22

17 **Cecil cut shapes out of paper. He placed 5 of them together on a table. The order of the shapes from left to right is**

trapezoid, square, rhombus, pentagon, triangle

Which diagram below shows the correct order of Cecile's shapes?

A

B

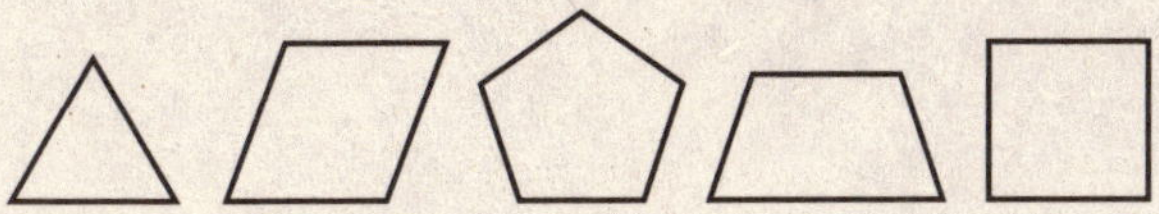

C

D

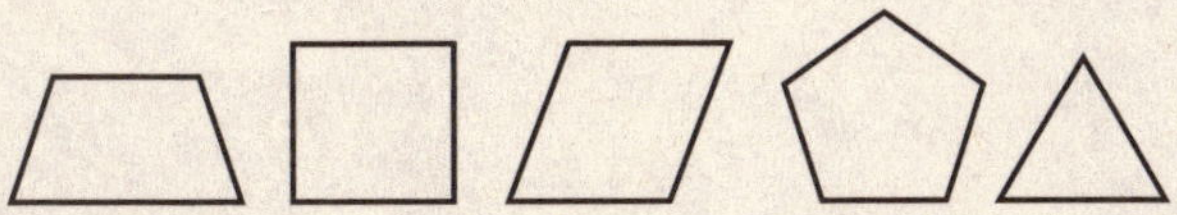

18 **Which of the following operations will result in a prime number?**

F $21 \div 7$

G $13 - 3$

H $17 + 5$

J 5×7

Go On

19 Jamilla plotted the rainfall in August for 4 towns on a number line. Which arrow points to the rainfall for Pleasantville, which was 9.8 centimeters?

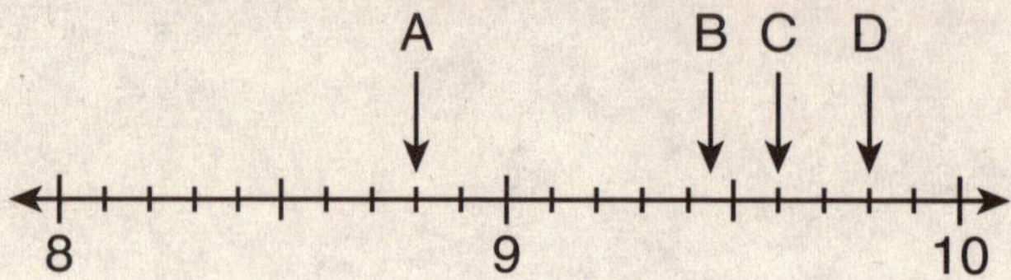

A A

B B

C C

D D

20 Which number will make this number sentence true?

$\square \div 2 < 12 \div 3$

F 6

G 8

H 10

J 12

21 **Shirley bought boxes of stationery. Each box contained 8 cards. How many cards could she have if she bought only complete packages?**

A 32

B 36

C 38

D 42

22 **A number cube showing the numbers 1, 2, 3, 4, 5, and 6 is rolled. What is the probability that a number less than 4 will be rolled?**

F $\frac{1}{4}$

G $\frac{3}{4}$

H $\frac{3}{6}$

J $\frac{4}{6}$

23 **Three students shot baskets on a basketball court. Zora shot more baskets than Nell, and Nell shot more baskets than Ramiro. Which statement below is true?**

A Ramiro shot more baskets than Zora.

B Ramiro shot the same number of baskets as Zora.

C Ramiro shot twice as many baskets as Zora.

D Ramiro shot fewer baskets than Zora.

Go On

24 Which expression below is equal to 9 × 6?

F 6 + 6 + 6 + 6 + 6 + 6

G 9 + 9 + 9 + 9 + 9 + 9

H 6 × 6 × 6 × 6 × 6 × 6

J 9 × 9 × 9 × 9 × 9 × 9

25 Leila used exactly 3 squares and 1 parallelogram to make a drawing. Which of the following could be Leila's drawing?

A

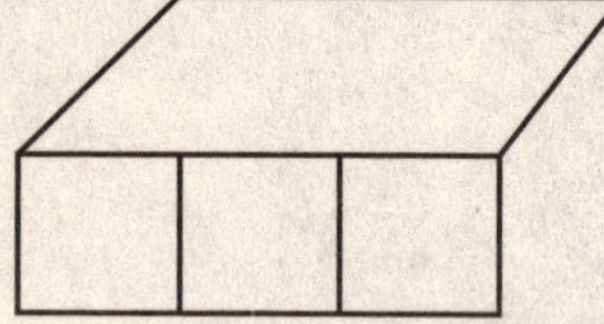

B

C

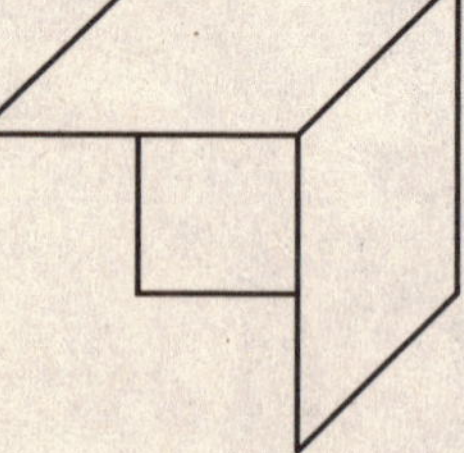

D

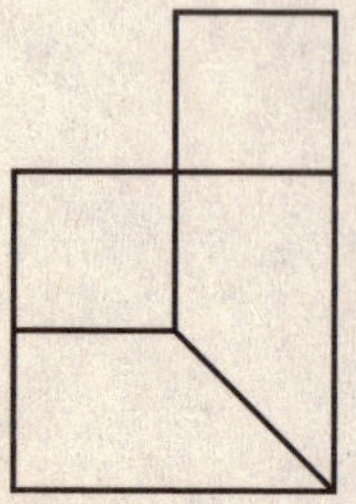

26 **Chet has 5 blocks. He has 1 white block, 2 gray blocks, and 2 black blocks. If the blocks are put in a bag, what is the probability of Chet randomly choosing a gray block?**

F 1 out of 2

G 1 out of 4

H 1 out of 5

J 2 out of 5

27 **For this problem, use your pattern blocks.**

Which sentence below is true about pattern block C?

A It is $\frac{1}{2}$ the size of pattern block D.

B It is $\frac{1}{4}$ the size of pattern block E.

C It is $\frac{1}{2}$ the size of pattern block B.

D It is $\frac{1}{4}$ the size of pattern block A.

28 **Kenyon is 57 inches tall. He is building a tower of 6-inch cubes. What is the maximum number of cubes that Kenyon can stack and still be taller than the tower?**

F 7

G 8

H 9

J 10

Go On

29 Mr. Moushey planted a vine near a grid he had made out of string. Look at the pattern the vine is making on the grid. If the vine continues its current pattern, which point on the grid will it reach?

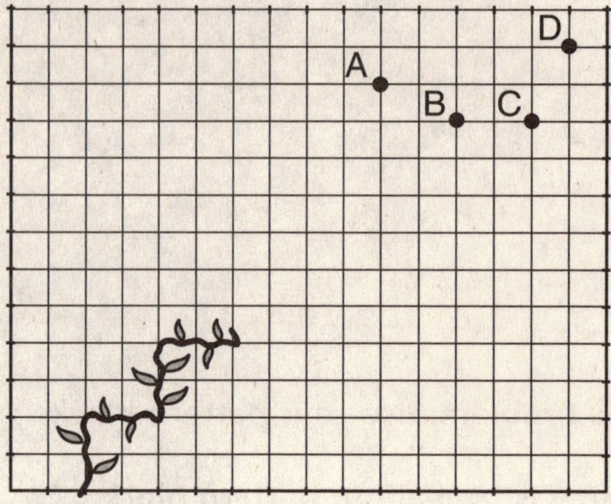

A A

B B

C C

D D

30 Tabitha labeled a set of 11 index cards with the numbers 1 through 11. If she places them face down on a table, shuffles them, and chooses a card at random, what are the chances that she will choose an even-numbered card?

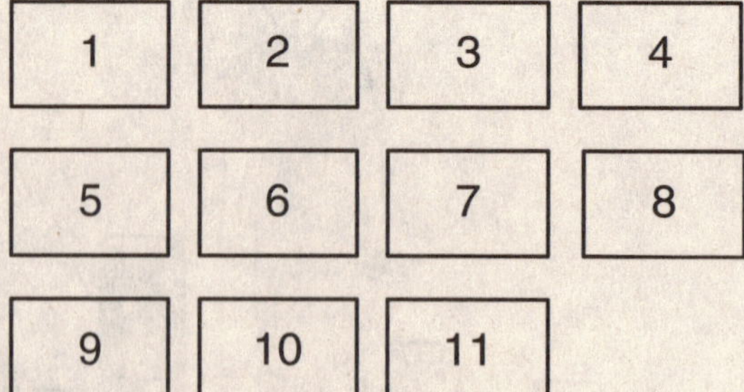

F 3 out of 11

G 4 out of 11

H 5 out of 11

J 6 out of 11

BOOK 2

TEST-TAKING TIPS

The suggestions below will help you do your best on the test.

- Take the time to read the directions in the Test Book carefully.
- If you do not understand the directions, ask your parent or teacher to explain them to you.
- You should read each question carefully, think about the answer, and then make your response.
- Show all of your work.
- Use only No. 2 pencils for the test.
- You may use tools such as your ruler, square counters, or pattern blocks to help you solve some of the test problems.

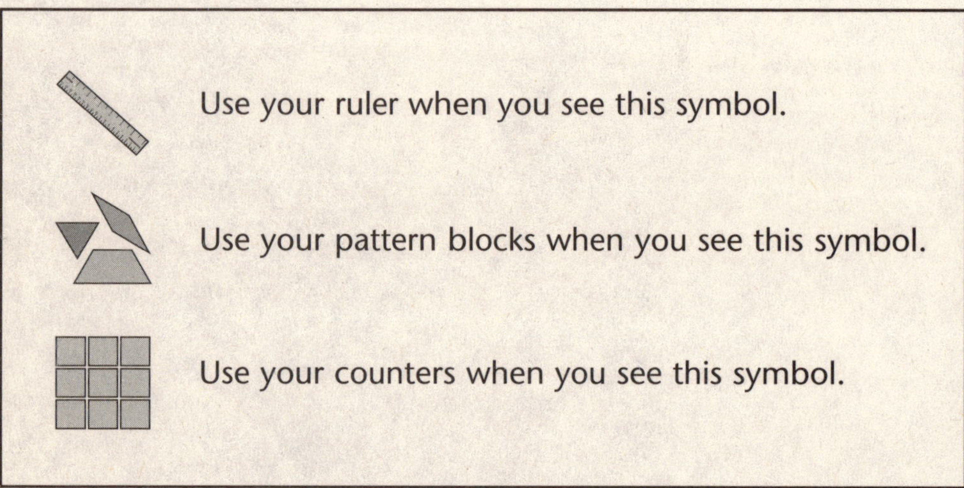

Session 2

31  **For this problem, use your ruler.**

Tiffany used cardboard to make a stencil for the capital letter T. A diagram of the stencil she used is shown below.

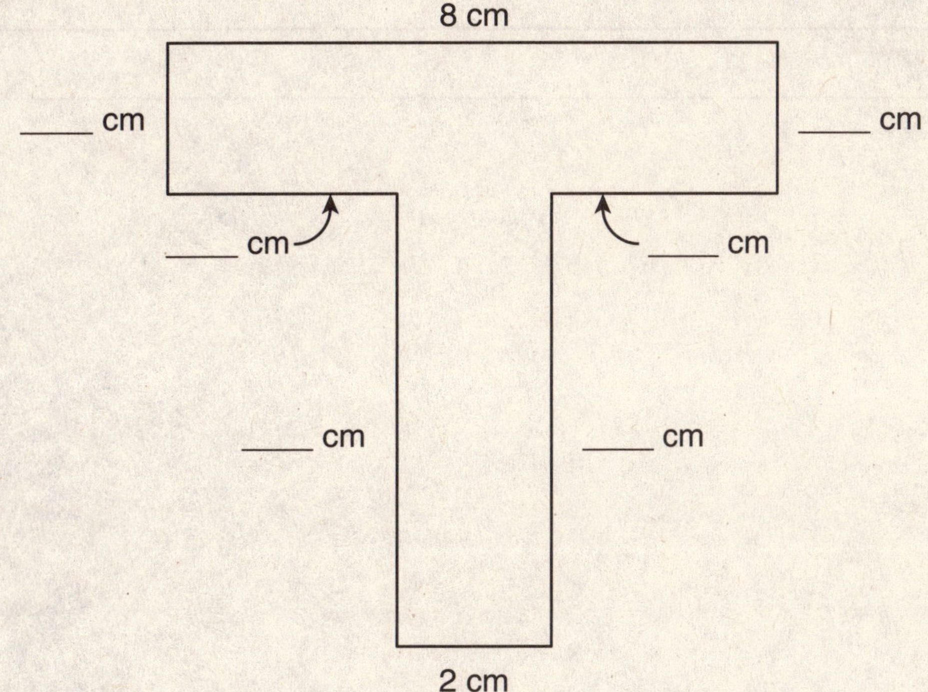

Part A

Measure each of the sides of Tiffany's stencil with your ruler. Write the missing measurements on the lines.

Part B

What is the perimeter of Tiffany's stencil?

Show your work.

Answer ______________________ **centimeters**

Go On

32 Look at the number sentence below.

$$9 \times (6 \times 14) = (9 \times 6) \times 14$$

Are both sides of this number sentence equal? Explain how you know the answer to this without performing the operations.

____

____

____

33 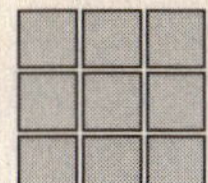 **For this problem, use your counters.**

Samuel took a bunch of 12 grapes from the fruit bowl. For breakfast, he ate $\frac{1}{3}$ of them. For lunch, he ate $\frac{1}{2}$ of the remaining grapes. How many total grapes did he eat for breakfast and lunch?

Show your work.

Answer ____________________ **grapes**

Go On

34 Look at the contents of the four bottles of water below.

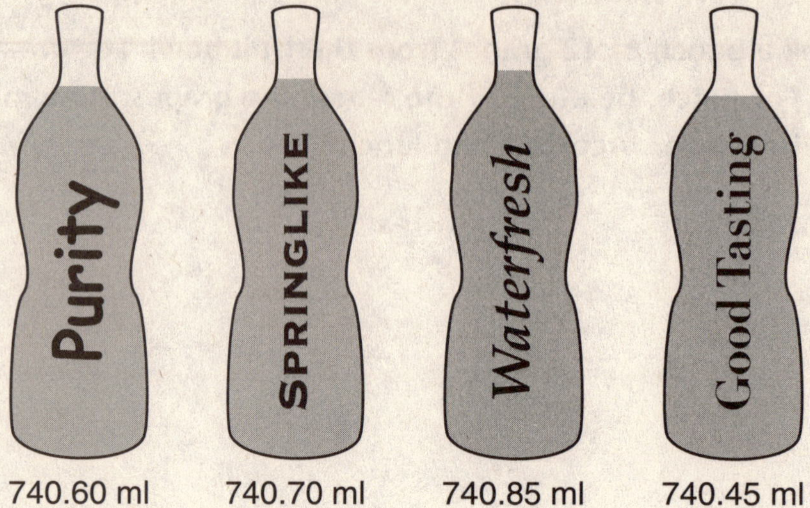

Part A

What is the name on the bottle that contains the amount closest to $740\frac{3}{4}$ milliliters?

Answer ____________________

Part B

Which bottle has more water than Good Tasting but less than Springlike?

Answer ____________________

35 The table below shows the number of goals the Saratoga Springs soccer team scored over 4 weeks.

Goals at Saratoga Springs Soccer Games

Week	Goals
1	6
2	4
3	1
4	8

Use the information from the table to make a bar graph on the grid below, showing all the soccer team's goals.

Remember to

- give the graph a title
- put labels on the axes
- graph the data you've been given
- make an appropriate scale

Go On

36 Mitchell wants to organize nails, screws, hooks, and other small pieces of hardware. At a yard sale, he finds a chest that has many drawers.

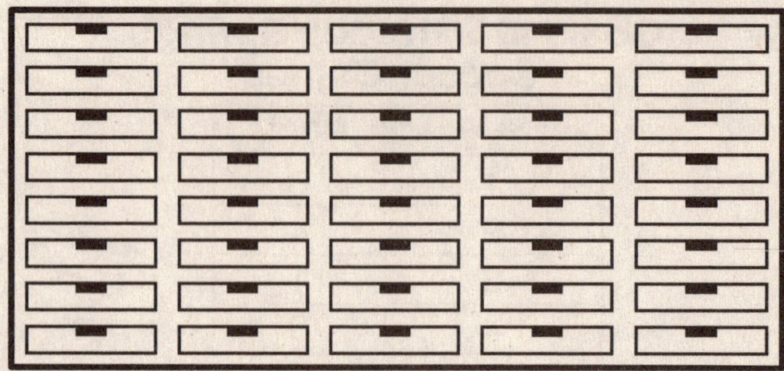

Write a multiplication equation that shows the number of drawers Mitchell has for storage.

Answer ______________________________

Use the same numbers in the equation above to write a division equation.

Answer ______________________________

37 LaTasha needs to save $28 to buy her mother a birthday present. She wants to save the same amount each week.

Part A

How much money does LaTasha need to save each week if she wants to buy the present in 4 weeks?

Answer ____________________ per week

How much money does LaTasha need to save each week if she wants to buy the present in 2 weeks?

Answer ____________________ per week

Part B

Could LaTasha save the exact same amount each week and buy the present in 3 weeks? Explain why or why not.

____

____

____

Go On

38 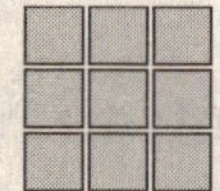 For this problem, use your counters.

Ms. Webster wants to tile her kitchen floor. Look at the floor shown below and use your counters to measure the open space. Each counter is equal to one tile.

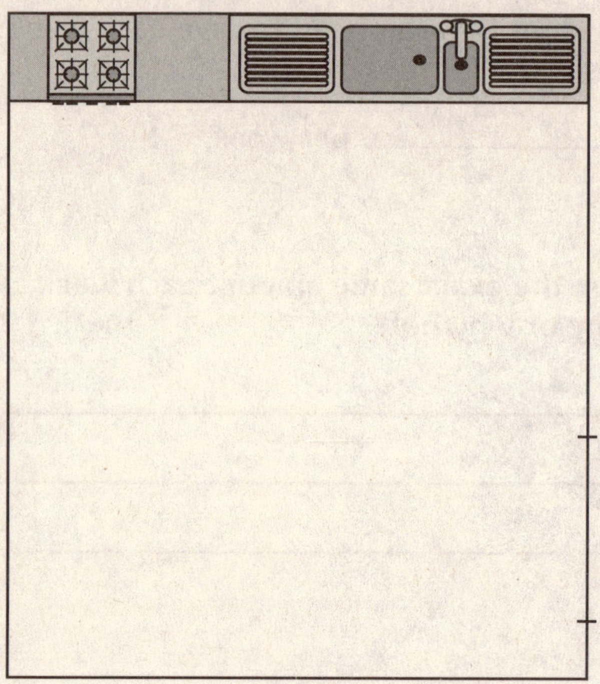

How many tiles will Ms. Webster need to tile her floor?

Answer ______________________ tiles

The area of each counter is 5 square units. What is the area of the floor in square units?

Answer ______________________ square units

39 There are 891 total students at Sam's school. His grade has 107 students. He wants to find out how many students are not in his grade.

Part A

How many students are not in Sam's grade?

Answer ______________________ students

Part B

The students who are not in Sam's grade are equally divided among the remaining 7 grades at his school. How many students are in each of the other grades?

Show your work.

Answer ______________________ students per grade

Session 3

40 For this problem, use your counters.

Dermott gave his mother 12 roses.

The roses were red, white, and yellow.

There were 4 red roses.

There were 3 times as many yellow roses as white roses.

Part A

How many yellow roses did Dermott give his mother?

Answer ____________________ yellow roses

Part B

How did you find your answer? Explain on the lines below.

____

____

____

Go On

41 **Beth Ann is playing a game with the spinners below. She does not want the arrow to land on a vowel.**

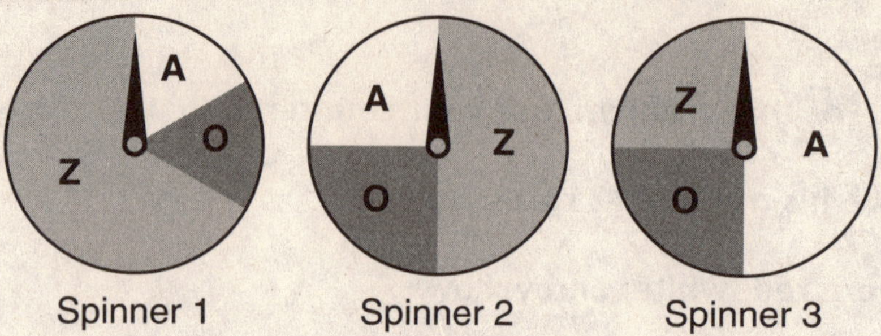

Which spinner will give Beth Ann the least probability of having the arrow point to a vowel?

Answer ____________________

Use the lines below to explain why you chose your answer.

____

____

____

42

Part A

Look at the diagram below to find the number pattern. Then draw the next set of circles in the series. Be sure to write the number of circles your drawing contains below it.

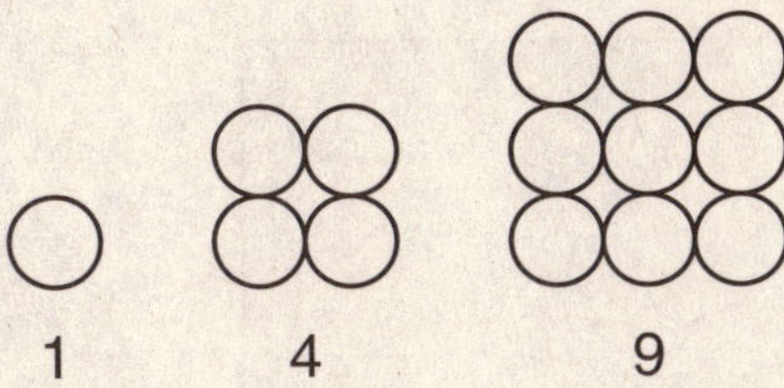

Part B

Look at the diagram below to find the number pattern. Then draw the missing set of circles in the series. Be sure to write the number of circles your drawing contains below it.

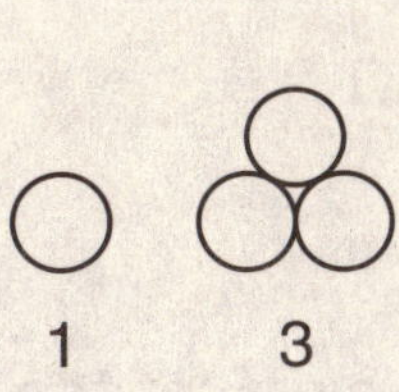

Go On

43 For this problem, use your counters.

Sheena cut a capital C from paper.

Part A

Sheena put 3 counters on top of the letter. How much of the letter is covered?

Answer ________________________

Part B

If Sheena had used 5 counters, what fraction of the letter would *not* be covered?

Answer ________________________

44 Look at the square shown on the graph below.

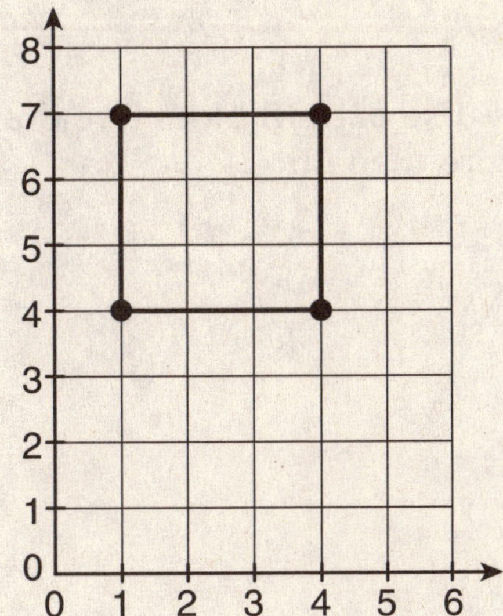

Part A

Double the length of the square on the graph. Connect the new coordinates. What type of shape have you made?

Answer ______________________

Part B

Write the ordered pairs of the points you plotted on the lines below.

(_____, _____)

(_____, _____)

Go On

45 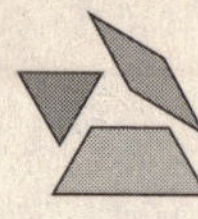 For this problem, use your pattern blocks.

Part A

Look at the shape below. Use pattern block E to draw the same shape. You may use the pattern block more than once.

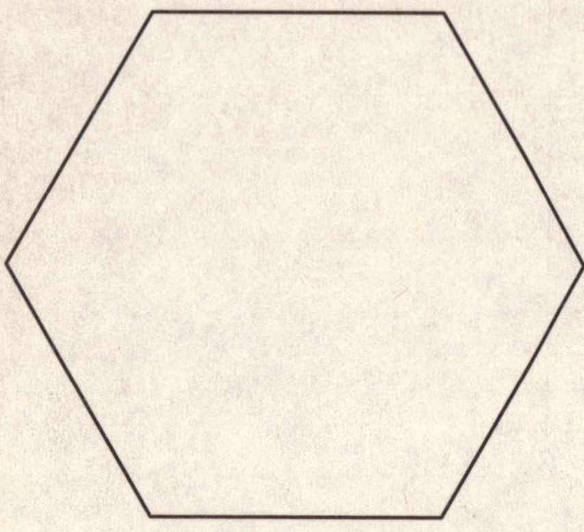

Part B

Use pattern block E to draw the same shape two times bigger.

How many pattern block Es did you use to draw the larger shape?

Answer ______________________ pattern block Es

46 Ms. Lane is painting a mural that will cover 3 walls of the gymnasium. Each wall will be painted with 22 athletes. If each athlete in the mural is painted holding 2 pieces of sports equipment, how many pieces of sports equipment will Ms. Lane paint?

Show your work.

Answer ______________________ pieces of sports equipment

Go On

47 Joacquin's sister's birthday is tomorrow. Joacquin wants to buy helium-filled balloons to give to her when she wakes up. He saw a sign at the grocery store that read:

Part A

How much would it cost to buy 10 balloons separately?

Show your work.

Answer $ ____________________

Part B

Joacquin has $4.40. What is the *greatest* number of balloons he can buy?

Show your work.

Answer ____________________ balloons

48 **Mr. Trad wants to put tar on his driveway. He drew the diagram below.**

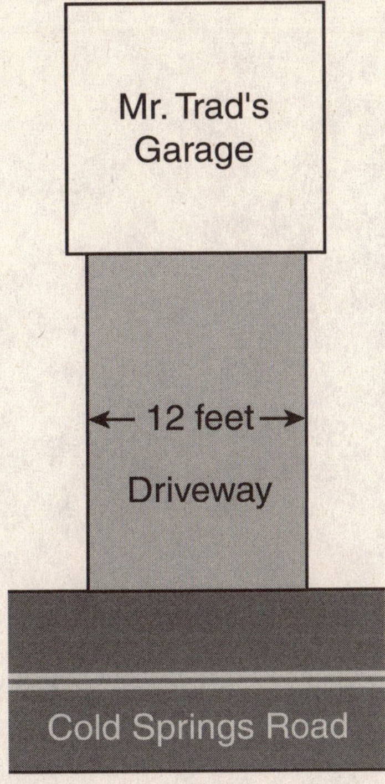

The perimeter of the driveway is 52 feet. What is the length of the driveway?

Show your work.

Answer ______________________ **feet**

Practice Test 2

BOOK 1

TEST-TAKING TIPS

The suggestions below will help you do your best on the test.

- Take the time to read the directions in the Test Book carefully.
- If you do not understand the directions, ask your parent or teacher to explain them to you.
- You should read each question carefully, think about the answer, and then make your response.
- Get rid of wrong answer choices when you don't know the right answer choice for a question.
- Use only No. 2 pencils for the test.
- You may use tools such as your ruler, square counters, or pattern blocks to help you solve some of the test problems.

Use your ruler when you see this symbol.

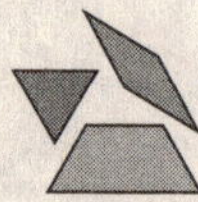

Use your pattern blocks when you see this symbol.

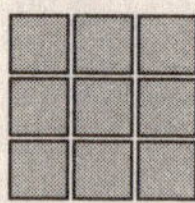

Use your counters when you see this symbol.

PRACTICE TEST 2 ANSWER SHEET

Completely darken bubbles with a No. 2 pencil. If you make a mistake, be sure to erase mark completely. Erase all stray marks.

1. YOUR NAME: ______________________________
(Print) Last First M.I.

SIGNATURE: ______________________________ DATE: ___/___/___

HOME ADDRESS: ______________________________
(Print) Number

____
City State Zip Code

PHONE NO.: ______________________________
(Print)

2. YOUR NAME

First 4 letters of last name				FIRST INIT	MID INIT
A	A	A	A	A	A
B	B	B	B	B	B
C	C	C	C	C	C
D	D	D	D	D	D
E	E	E	E	E	E
F	F	F	F	F	F
G	G	G	G	G	G
H	H	H	H	H	H
I	I	I	I	I	I
J	J	J	J	J	J
K	K	K	K	K	K
L	L	L	L	L	L
M	M	M	M	M	M
N	N	N	N	N	N
O	O	O	O	O	O
P	P	P	P	P	P
Q	Q	Q	Q	Q	Q
R	R	R	R	R	R
S	S	S	S	S	S
T	T	T	T	T	T
U	U	U	U	U	U
V	V	V	V	V	V
W	W	W	W	W	W
X	X	X	X	X	X
Y	Y	Y	Y	Y	Y
Z	Z	Z	Z	Z	Z

3. DATE OF BIRTH

Month	Day		Year			
JAN						
FEB						
MAR	0	0	0	0	0	0
APR	1	1	1	1	1	1
MAY	2	2	2	2	2	2
JUN	3	3	3	3	3	3
JUL		4	4	4	4	4
AUG		5	5	5	5	5
SEP		6	6	6	6	6
OCT		7	7	7	7	7
NOV		8	8	8	8	8
DEC		9	9	9	9	9

4. SEX

MALE

FEMALE

The Princeton Review

Practice Test 2

Session 1

Sample A (A) (B) (C) (D)

Sample B (F) (G) (H) (J)

Sample C (A) (B) (C) (D)

1. (A) (B) (C) (D)	11. (A) (B) (C) (D)	21. (A) (B) (C) (D)
2. (F) (G) (H) (J)	12. (F) (G) (H) (J)	22. (F) (G) (H) (J)
3. (A) (B) (C) (D)	13. (A) (B) (C) (D)	23. (A) (B) (C) (D)
4. (F) (G) (H) (J)	14. (F) (G) (H) (J)	24. (F) (G) (H) (J)
5. (A) (B) (C) (D)	15. (A) (B) (C) (D)	25. (A) (B) (C) (D)
6. (F) (G) (H) (J)	16. (F) (G) (H) (J)	26. (F) (G) (H) (J)
7. (A) (B) (C) (D)	17. (A) (B) (C) (D)	27. (A) (B) (C) (D)
8. (F) (G) (H) (J)	18. (F) (G) (H) (J)	28. (F) (G) (H) (J)
9. (A) (B) (C) (D)	19. (A) (B) (C) (D)	29. (A) (B) (C) (D)
10. (F) (G) (H) (J)	20. (F) (G) (H) (J)	30. (F) (G) (H) (J)

Write answers to Questions 31–48 in the space provided.

Session 1

Sample A

```
  436
+  29
```

A 445

B 455

C 465

D 565

Sample B

Which page will be next?

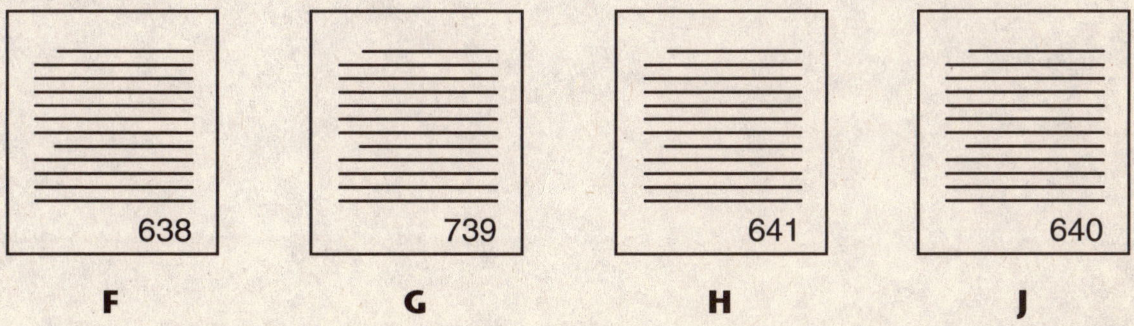

F **G** **H** **J**

Go On

Sample C

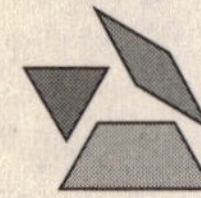

For this problem, use your pattern blocks.

How many of pattern block C would be needed to cover exactly $\frac{1}{2}$ of the shape above?

A 2

B 3

C 4

D 5

1

$$\begin{array}{r} 408 \\ -\ 36 \\ \hline \end{array}$$

A 362

B 372

C 444

D 472

2

$$\begin{array}{r} 53 \\ \times\ 6 \\ \hline \end{array}$$

F 308

G 309

H 318

J 319

Go On

3 **Which of the following expressions is *not* equal to 6 × 4?**

A 4 + 4 + 4 + 4 + 4 + 4

B 4 × 6

C 6 + 6 + 6 + 6

D 6 × 6 × 6 × 6

4

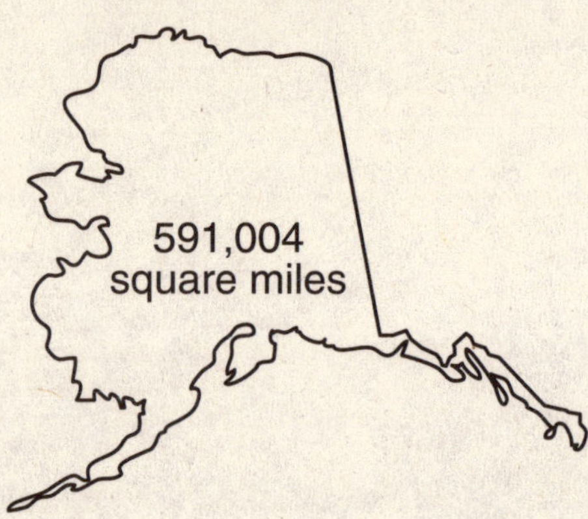

Which of the following is *not* true about the number 591,004?

F The 4 is in the thousands place.

G An odd digit is in the hundred thousands place.

H The 9 is in the ten thousands place.

J The 1 has a value of 1,000.

5 **Which of the following is the most likely height of a telephone pole?**

A 10 centimeters

B 10 decimeters

C 10 meters

D 10 kilometers

6 **Kyra has 9 rolls of pennies. Each roll contains 50 pennies. How many pennies does Kyra have?**

F 140

G 450

H 590

J 900

7 **Which fraction is *less* than $\frac{1}{2}$?**

A $\frac{2}{7}$

B $\frac{3}{4}$

C $\frac{4}{5}$

D $\frac{5}{6}$

Go On

8 **Ms. Matthews drew these shapes on the board.**

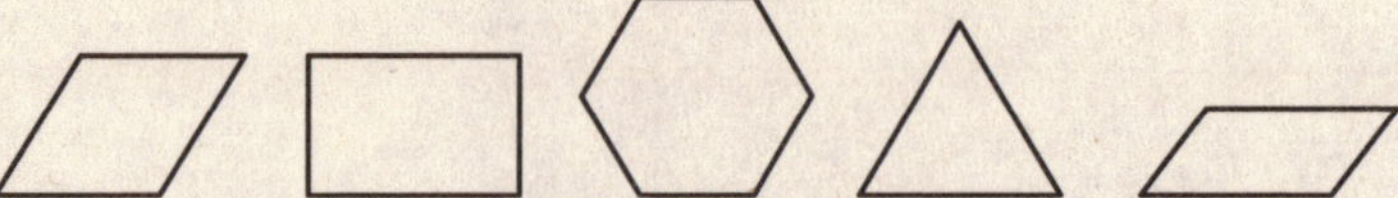

Which of the following lists the shapes in the correct order?

F rhombus, parallelogram, hexagon, rectangle, triangle

G hexagon, rhombus, parallelogram, triangle, rectangle

H triangle, rectangle, rhombus, hexagon, parallelogram

J rhombus, rectangle, hexagon, triangle, parallelogram

9 **Which number will make this number sentence true?**

$\square + 12 + 7 < 3 + 8 + 11$

A 2

B 3

C 4

D 5

10 **There were 100 points available on a test. Newton earned 85 points on the test. What percent of the 100 points did Newton earn?**

F 0.85%

G 8.5%

H 85%

J 850%

11 **What fraction of the cup below is empty?**

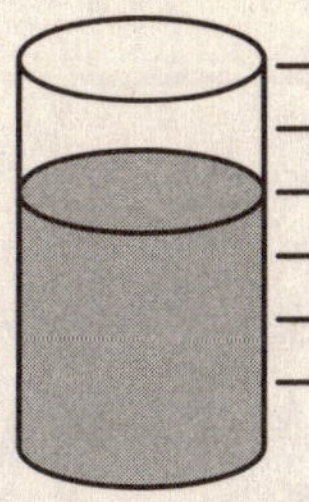

A $\frac{1}{6}$

B $\frac{2}{6}$

C $\frac{3}{6}$

D $\frac{4}{6}$

12 **Mr. Williams bought boxes of colored chalk for summer camp. Each box contained 12 pieces of chalk. If Mr. Williams bought only complete packages, how many pieces of chalk could he have bought?**

F 64

G 68

H 70

J 72

13 **Darryl slept for 8 hours and 45 minutes. Which of the following could be the time that Darryl slept?**

A 10:15 P.M. to 8:00 A.M.

B 10:45 P.M. to 8:00 A.M.

C 11:15 P.M. to 8:00 A.M.

D 11:45 P.M. to 8:00 A.M.

Go On

14 **Look at the bags of marbles below. Which bag would give you a 1 out of 3 probability of choosing a gray marble?**

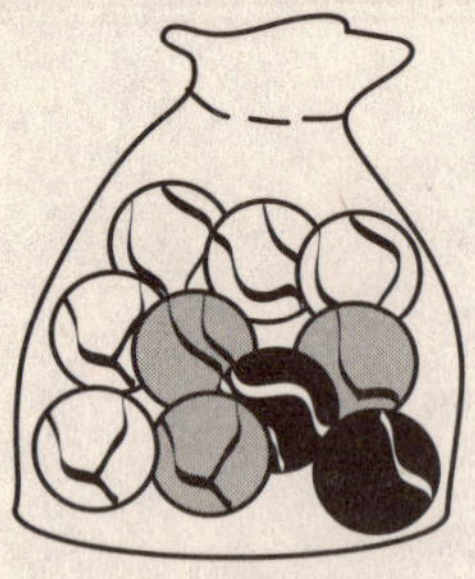
Pat's Marbles

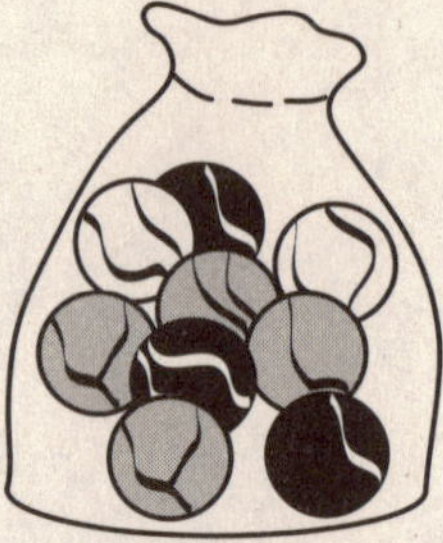
Mel's Marbles

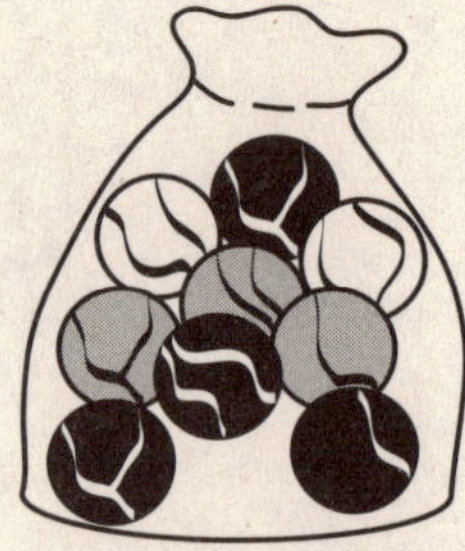
Rafi's Marbles

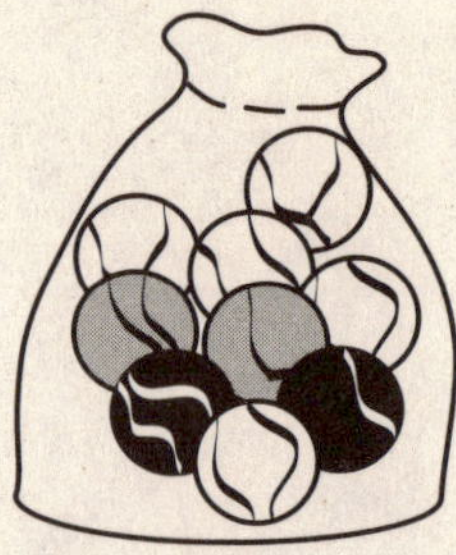
Carol's Marbles

F Pat's marbles

G Mel's marbles

H Rafi's marbles

J Carol's marbles

15 **Look at the pattern below. Which number belongs in the box?**

☐, 42, 49, 56, 63

A 27

B 35

C 37

D 40

16 **For this problem, use your pattern blocks.**

Which sentence below is true?

F Pattern block A is 3 times the size of pattern block C.

G Pattern block A is 2 times the size of pattern block E.

H Pattern block A is $\frac{1}{2}$ the size of pattern block B.

J Pattern block A is $\frac{1}{2}$ the size of pattern block E.

17 **Grainne weighed 3 bags of apples at the store. She found that R weighed less than T, and S weighed less than R.**

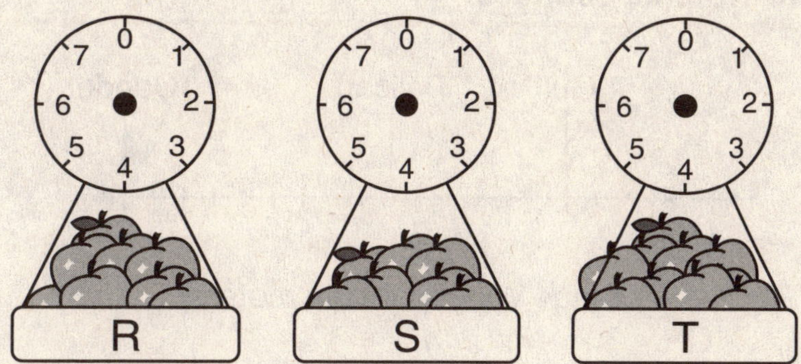

Which statement below is true?

A T weighs more than S.

B T weighs the same as S.

C T weighs less than S.

D T weighs twice as much as S.

18 **Which statement is true about the figure below?**

F It has no parallel lines.

G It has exactly 1 pair of parallel lines.

H It has exactly 2 pairs of parallel lines.

J It is a trapezoid.

Go On

19 The number line below shows the average wingspans in meters of different birds. Find a possible wingspan of an eagle if its wingspan is greater than a pelican's but less than a condor's.

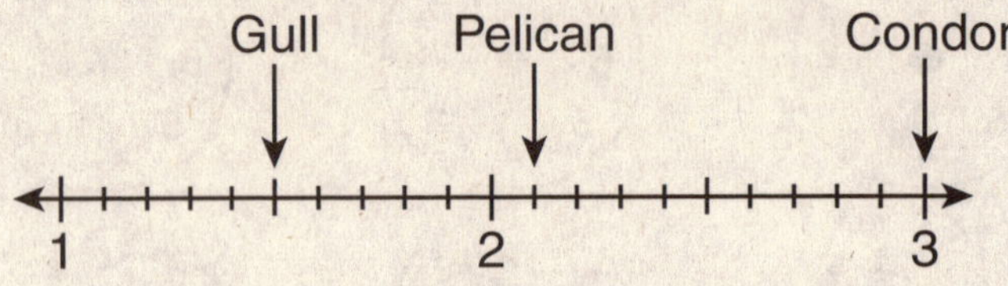

A 1.3 meters

B 1.8 meters

C 2.1 meters

D 2.5 meters

20 Brittany used 2 triangles and 1 trapezoid to make a design. Which drawing below is the one that Brittany made?

F 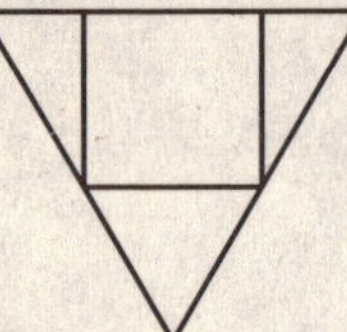

G 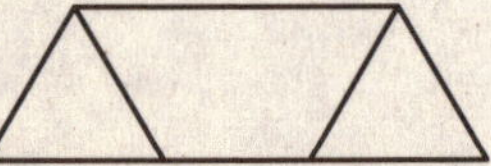

H 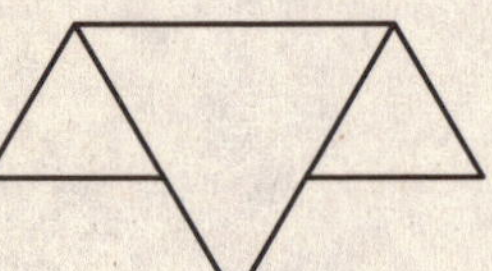

J

21 There were 595 students at Dixie Elementary School last year. During the summer, 19 students moved away. If no new students enter the school, which expression shows the number of students at the school this year?

A $595 - 19$

B $595 + 19$

C 595×19

D $595 \div 19$

22 Tia is playing a game with a spinner.

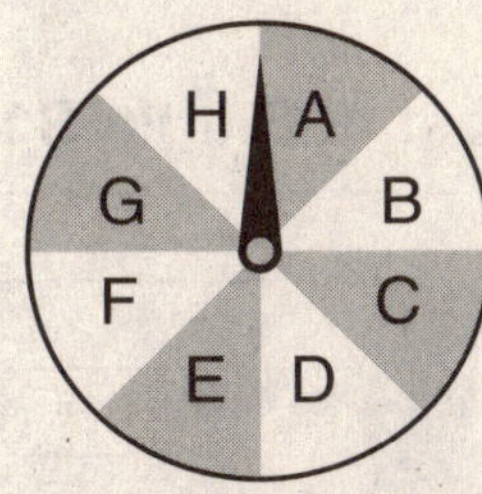

What is the probability that Tia's spinner will land on a vowel?

F $\frac{1}{8}$

G $\frac{2}{8}$

H $\frac{5}{8}$

J $\frac{6}{8}$

23 Jasmine had 35 cans to take to the recycling center. She put them into boxes that held 8 cans each. If she completely filled each box before filling the next one, how many cans were in the last box?

A 7

B 6

C 5

D 3

Go On

24 Mr. Ricardo has 10 pieces of wire that are each 5.9 centimeters long. About how many centimeters of wire does he have in all?

F 40

G 50

H 60

J 70

25 Devon compared the cost of single-scoop ice cream cones in her town. The graph below shows the prices of single-scoop ice cream cones at different shops.

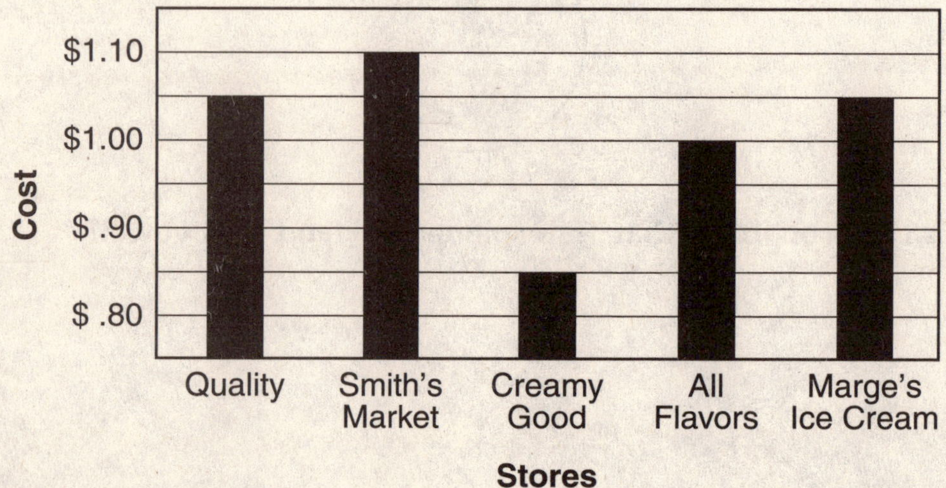

How much did All Flavors charge for a single-scoop cone?

A $0.95

B $1.00

C $1.05

D $1.10

26 **Which of the following will result in an even sum?**

F even number + odd number

G even number + even number + odd number

H odd number + odd number

J odd number + odd number + odd number

27 **Stan and Jerry are following directions to get around a new town. They were told to continue walking in the pattern shown below.**

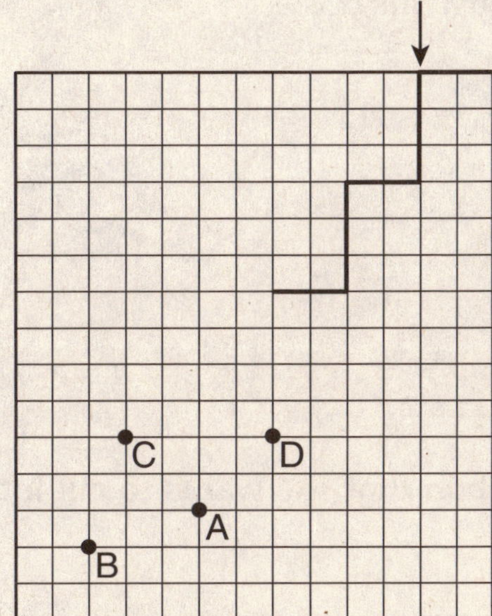

Which location will they reach if they continue walking in this pattern?

A A

B B

C C

D D

Go On

28 Friendly's Pet Store has 9 gerbils in the store. There are 4 fewer puppies than gerbils. There are 6 more kittens than puppies. How many kittens are there?

F 6

G 9

H 10

J 11

29 Jonathan has a drawer with only T-shirts in it. There are 2 white, 3 red, and 2 blue T-shirts. What is the probability that Jonathan will reach into the drawer and randomly pick a red T-shirt?

A $\frac{1}{3}$

B $\frac{3}{6}$

C $\frac{3}{7}$

D $\frac{4}{7}$

30 Ivana has a piece of ribbon that she wants to cut into equal pieces. Each piece will be 6 inches long.

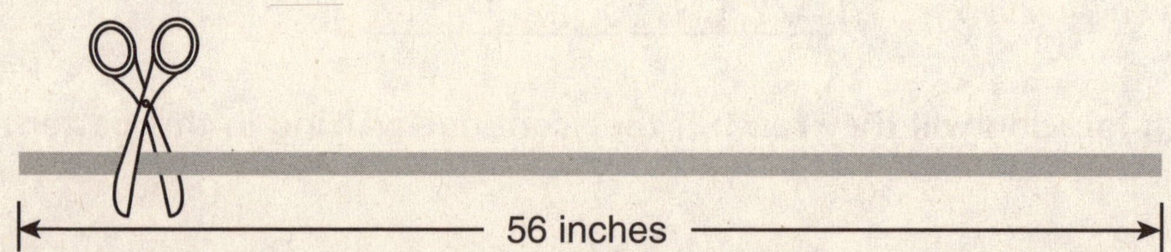

What is the greatest number of equal pieces that Ivana can cut from the ribbon?

F 7

G 8

H 9

J 10

BOOK 2

Test-Taking Tips

The suggestions below will help you do your best on the test.

- Take the time to read the directions in the Test Book carefully.
- If you do not understand the directions, ask your parent or teacher to explain them to you.
- You should read each question carefully, think about the answer, and then make your response.
- Show all of your work.
- Use only No. 2 pencils for the test.
- You may use tools such as your ruler, square counters, or pattern blocks to help you solve some of the test problems.

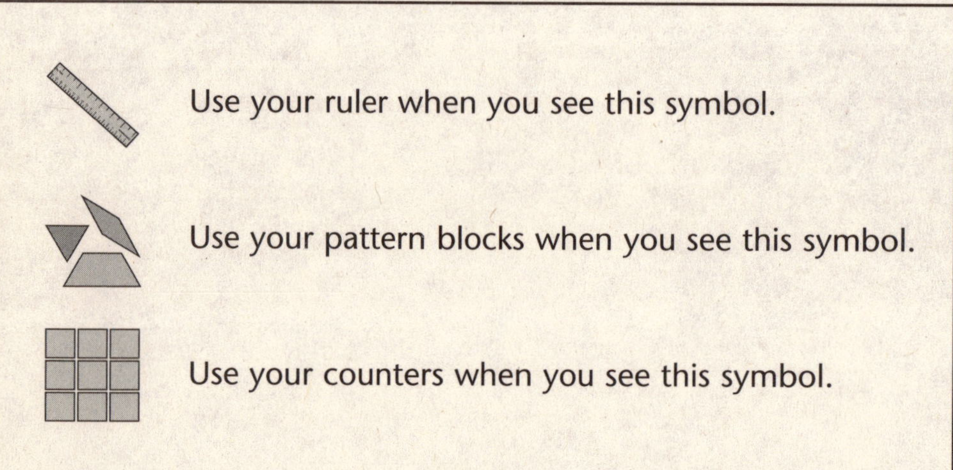

Session 2

31 **Peg and Lacy are playing a number game, and the first number on one of the game cards is missing.**

72 ✪ 8 = 9

44 ✪ 11 = 4

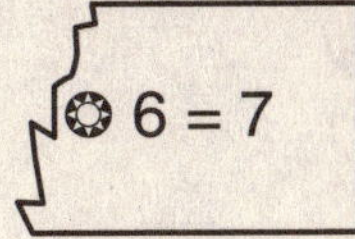

What operation does the symbol stand for?

Answer ______________________

What number is missing from the torn card?

Answer ______________________

Go On

32  For this problem, use your counters.

Mr. Pauling has a job selling houses. He sold 2 houses in January. He sold twice that number of houses in February. In March, Mr. Pauling sold $1\frac{1}{2}$ times the number of houses he sold in February. What is the total number of houses that Mr. Pauling sold?

Show your work.

Answer ______________________ houses

33 Randy has a fish tank that holds 10 gallons of water. Because of evaporation, the water in the tank decreases over time. Randy has to add 8 ounces of water each week to keep the tank full.

Part A

How much total water did Randy add at the end of 9 weeks?

Answer ____________________ ounces

How much total water did Randy add at the end of 14 weeks?

Answer ____________________ ounces

Part B

If Randy did not add water to the fish tank each week, would the tank have lost more than 1 gallon of water after 18 weeks? (Hint: 1 gallon of water is equal to 128 ounces.) Why or why not?

____

____

____

Go On

34 Look at the number sentence below.

$$154 + 23 + 78 + 12 = 12 + 78 + 23 + 154$$

Is this number sentence true? Explain how you know without solving it.

____

____

____

35 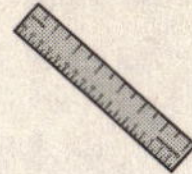 **For this problem, use your ruler.**

Consuela decided to make gift boxes for her friends. She designed a triangular shape for the boxes and drew the pattern below.

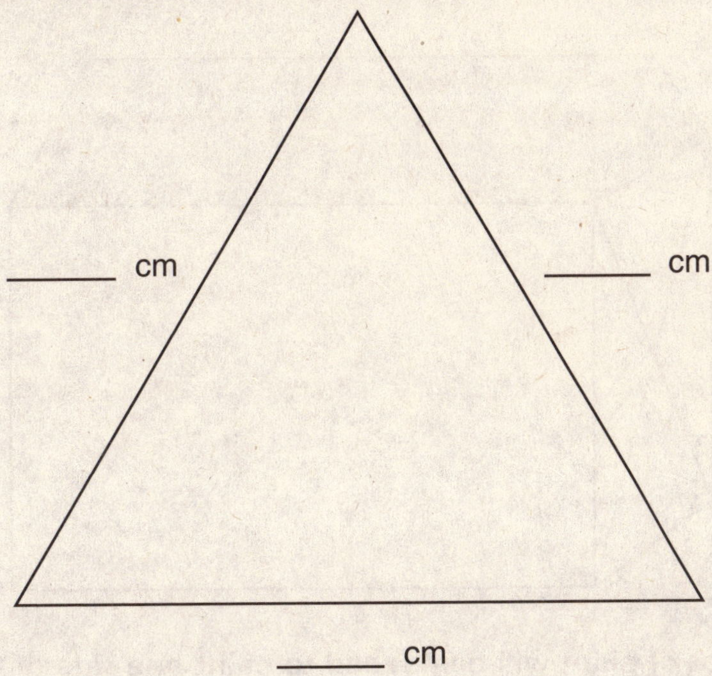

Part A

Use your ruler to measure each side of Consuela's box. Write the measurements on the lines next to the diagram.

Part B

Consuela wants to add a decorative trim to her boxes and needs to calculate the distance around each box. What is the perimeter of the box?

Show your work.

Answer _______________ **centimeters**

Go On

36 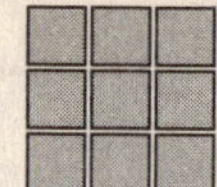 **For this problem, use your counters.**

The playground at Roland's school has a shed for storing sports equipment. Roland and his friends have volunteered to repaint the shed, shown below. One cup of paint will cover the area of one of your counters.

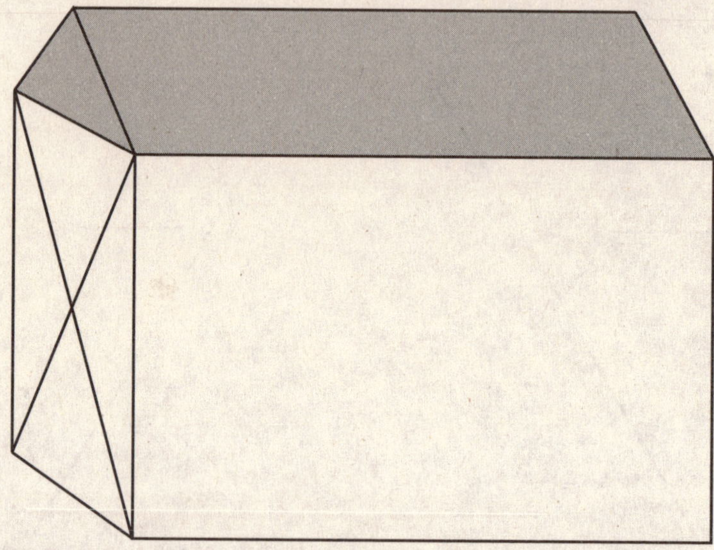

How many cups of paint will be needed to paint one side of the shed?

Answer ____________________ **cups**

One counter has an area of 16 square units. Calculate the area, in square units, of one side of the shed.

Answer ____________________ **square units**

37 As an experiment, Pacquin weighed four boxes of raisins to see if they contained the same amount of raisins. The table below shows the results of his experiment.

Weight of Raisins

Box Number	Weight (ounces)
1	0.80
2	0.52
3	0.67
4	1.02

Part A

Which box of raisins weighs closest to $\frac{7}{10}$ ounce?

Answer ______________________

Part B

Which box weighed more than Box 3 but less than Box 4?

Answer ______________________

Go On

38 The tallest redwood tree measures 368 feet. Another redwood tree measures 172 feet shorter than the tallest redwood tree.

Part A

What is the height of the shorter redwood tree?

Answer ______________________ feet

Part B

A maple tree in Smith Park is 4 times shorter than the tallest redwood. If an average maple tree grows to 100 feet, how many feet shorter than average is the tree in Smith Park?

Show your work.

Answer ______________________ feet

39 Ms. Cianci teaches dance classes after school. She took a survey to find out what type of dance her students would like to learn. The table below lists the results of her survey.

Dance Survey

Type of Dance	Number of Students
Tap	6
Ballet	11
Modern	9
Step	4
Jazz	8

Make a bar graph on the grid below that uses the information from the table.

Remember to

- give the graph a title
- put labels on the axes
- graph the data you've been given
- make an appropriate scale

Session 3

40 J'Vante made the squares shown below.

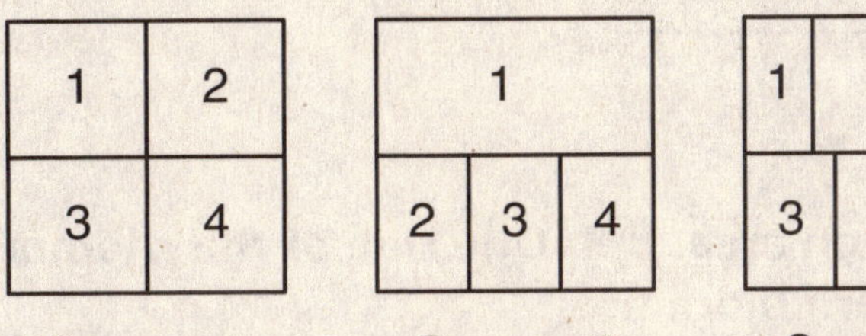

J'Vante tosses a coin onto each of the squares. Which square has the greatest probability of having the coin land on an even number?

Answer ____________________

Use the lines below to explain why the square you chose will have the greatest probability of a coin landing on an even number.

____

____

____

Go On

41 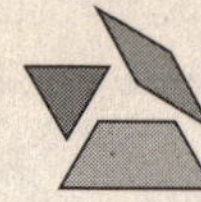 For this problem, use your pattern blocks.

Part A

If pattern block C has an area of 1, which pattern block has an area of 2?

Answer ______________________

Part B

Draw a rhombus with an area of 8. One side of the rhombus will be exactly 2 times the side of pattern block C.

42 **Niles needs to buy poster paper for a science project. At the store, he finds two ways of buying the poster paper.**

Part A

How much less will Niles spend if he buys 4 sheets of poster paper instead of a value pack?

Show your work.

***Answer* $**____________________

Part B

Niles's sister needs to buy 8 sheets of paper. What is the *least* amount of money she can spend?

Show your work.

***Answer* $**____________________

Go On

43 For this problem, use your counters.

Look at the rectangle below.

Part A

Cover $\frac{3}{4}$ of the rectangle with your counters. How many counters did you use?

Answer ______________ counters

Part B

If you covered the rectangle with 6 counters, what fraction of it would be covered?

Answer ______________

44

Part A

Find the pattern in the table below.

1	2
3	6
5	10
7	14

What numbers would go into the sixth row?

______, ______

Part B

Find the pattern in the table below.

18	6
15	5
9	3
6	2
3	1

What numbers would go into the third row?

______, ______

Go On

45 Ms. Kipfer made stained-glass sun catchers in 3 designs. She sold her sun catchers at 4 different stores. If she sent each store 15 sun catchers of each design, how many sun catchers did she send altogether?

Show your work.

Answer ______________________ sun catchers

46 Look at the figure shown on the graph below.

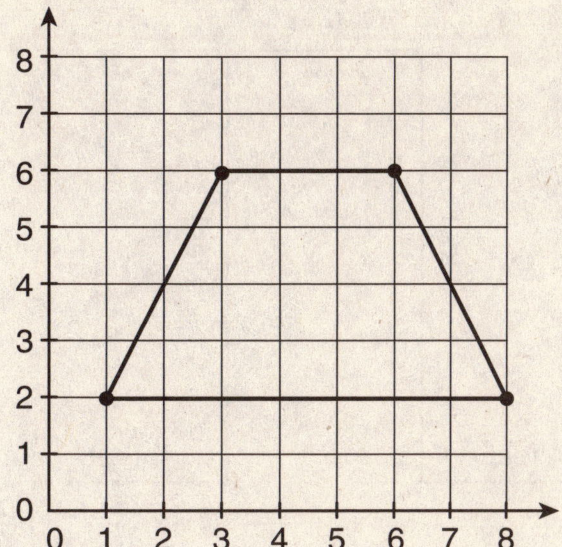

Part A

What is the geometric shape on the graph?

Answer ______________________________

Part B

Write the ordered pairs of the corner points of the figure on the lines below.

(_____, _____)

(_____, _____)

(_____, _____)

(_____, _____)

Go On

47 **Deva drew a diagram of her garden as shown below.**

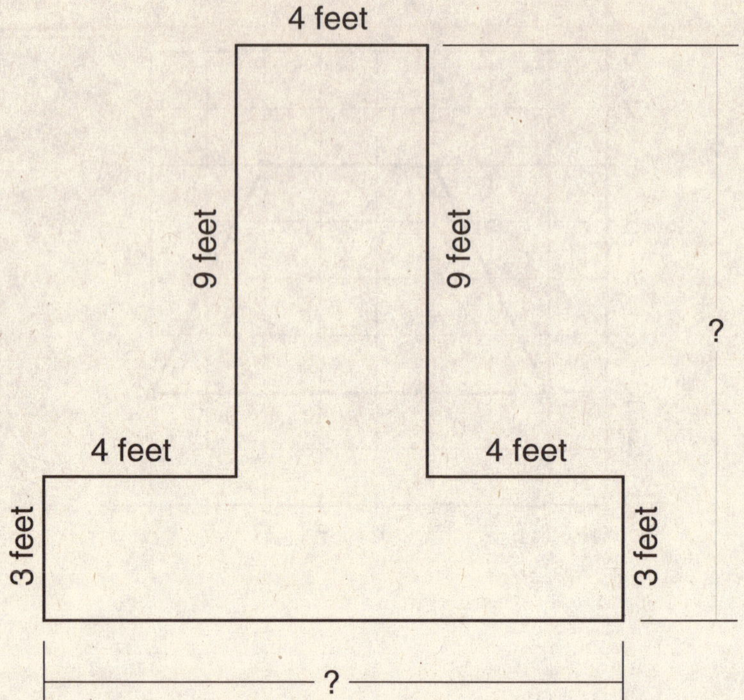

Part A

Find the total length and width of the entire garden.

Length ______________________ **feet**

Width ______________________ **feet**

Part B

Deva is going to put a fence around her garden. Find the perimeter of the garden.

Show your work.

Answer ______________________ **feet**

48  For this problem, use your counters.

Raphael had 14 colored markers.

- Each marker was either red, yellow, green, or blue.
- There were 3 yellow markers and 2 green markers.
- There were half as many red markers as blue markers.

Part A

How many red markers did Raphael have?

Answer ____________________ red markers

Part B

How did you find your answer? Explain on the lines below.

____

____

____

Answers and Explanations for Practice Test 1

SESSION 1

Sample A **B** When you subtract 38 from 866, you have to borrow to find the correct answer of 828. When borrowing, make sure that you cross off the original number and write the new number neatly. It's easy to make a mistake when a number isn't written clearly. If you need to practice borrowing in subtraction, review Mile 6.

Sample B **H** To find the number that would be next, find the pattern in the series of numbers. The numbers are increasing, so you will be adding to find the next number. Add 0.5 to 7.5 to get 8. Add 0.5 to 8 to get 8.5. Add 0.5 to 8.5 to get 9. Adding 0.5 to each of the numbers is the pattern. Add 0.5 to 9 to find the next number. It is 9.5, or answer choice **H.**

Sample C **D** The perimeter of a rectangle is the sum of all of its sides. Use your ruler carefully to measure each side. $2 + 2 + 1 + 1 = 6$. The perimeter is 6 inches.

1. **C** $18 \times 7 = 126$. In this multiplication problem, remember to carry properly. If you thought the answer was choice **B,** 116, then you didn't carry the 5 correctly. You can practice more of these problems in Mile 7. If you multiplied incorrectly, slow down. You can avoid mistakes if you take more time to think about your answer.

2. **G** $266 + 28 = 294$. If you thought the answer was choice **F,** 284, then you didn't carry the 1. Don't forget to carry! If you had trouble with the problem, go to Mile 5 to practice. If you made an addition error, slow down. Taking time to solve problems that look easy can help you avoid mistakes.

3. **C** One way to solve this problem is to find the number of minutes until the beginning of the next hour. The amount of time from 8:15 A.M. to 9:00 A.M. is 45 minutes. Then find the number of hours from 9:00 to 3:00. That's 6 hours. The answer is 6 hours, 45 minutes.

4. **J** To answer this question, get rid of the wrong answer choices. The first answer choice is wrong because the 0 is in the ones place. Answer choice **G** is wrong because the value of the 1 is 10,000. The number in the hundreds place is odd. Get rid of answer choice **H.** That leaves answer choice **J,** which is correct because the 9 is in the thousands place.

5. **C** For a question like this, carefully write down the information as it's given to you. Cara has 6 blue marbles and 2 fewer yellow marbles. $6 - 2 = 4$ yellow marbles. She has 5 more red marbles than yellow marbles. $5 + 4 = 9$ red marbles. Answer **C** is correct.

6. **G** If you don't know the answer to this question, get rid of the wrong answer choices to help you. Answer choice **F** is wrong because 12 grams is about the weight of 12 paper clips. Answer choice **G** sounds about right, but look at the other choices. Twelve pounds is heavier than some bags of potatoes. Get rid of answer choice **H.** Twelve kilograms is about the weight of a small bike. Get rid of it. You're left with 12 ounces, which is answer choice **G.**

7. **A** It might help you to divide a circle into equal parts to find the size of each fraction. For $\frac{1}{4}$, draw a circle and divide it into 4 equal sections. For $\frac{1}{3}$, draw another circle and divide it into 3 equal sections. You can see that if a circle is divided into thirds, each third is larger than a section of a circle divided into fourths. So $\frac{1}{3}$ is more than $\frac{1}{4}$. If you want to check the other answer choices, divide a circle into 5 equal parts, 6 equal parts, and 8 equal parts. You will see that $\frac{1}{4}$ is larger than $\frac{1}{5}$, $\frac{1}{6}$, and $\frac{1}{8}$. None of those could be the correct answer.

8. **H** Shawn has 7 parts of 10 dollars, or $\frac{7}{10}$. Think about a grid that is divided into 10 parts. Shade 7 parts. 70% of the grid is shaded. $\frac{7}{10} = 70\%$. Shawn has 70% of the cost of a CD, or answer choice **H.**

9. **D** You can get rid of the wrong answer choices to help you solve this question. Look at the bar graph for Sunday. Forty cars crossed the bridge that day. The question asks for a day when *less* than 40 cars crossed. Answer choice **A** is wrong. Get rid of it. On Monday, 30 cars crossed the bridge. The question asks for a day when *more* than 30 cars crossed the bridge. Answer choice **B** is wrong. Get rid of it. On Friday, 25 cars crossed the bridge. That's another wrong answer. Get rid of that one too. Look! All that's left is answer choice **D.** It must be right. The bar graph says 35 cars crossed the bridge on Saturday. It is the correct answer!

10. **G** This is an estimation question. You don't have to figure it out exactly. You can round down 710 to about 700. Ms. Hammet wants to put the cookies into 10 boxes, so you would divide 700 by 10. That equals 70. Answer choice **G** is correct.

11. **A** Parallel lines are lines that will never intersect. You want to find the figure that has *no* parallel lines. Answer choice **A** is a right triangle. The lines are intersecting, so they aren't parallel. That looks like a good answer. Just to make sure, look at answer choice **B.** It's a square. A square has 2 pairs of parallel lines, so it's the wrong answer. The trapezoid has 1 pair of parallel lines. It's wrong too. The parallelogram also has 2 pairs of parallel lines, so it's also wrong. The only figure that does not have any parallel lines is the triangle, which is answer choice **A.**

12. **G** You want to find out how many stickers Susan should give equally to 8 people. She has 96 stickers. Divide 96 by 8 to find out how many each person should get. $96 \div 8 = 12$. Susan will give each person 12 stickers, which is answer choice **G**.

13. **D** The problem $35 \times 12 = 420$ is multiplication. When you check to see that the answer to a multiplication question is correct, you use division. You would divide the product by one of the factors. In this case, the product is 420 and the factors are 35 and 12. The only answer choice that has the product divided by one of the factors is answer choice **D,** $420 \div 12$.

14 **F** You need to solve this problem in steps. First, multiply 5 by 7 to see how many tapes Jeremy can hold in his travel cases. $5 \times 7 = 35$. Jeremy can hold 35 tapes in his travel cases. Next, subtract 32 from 35 to see how many more tapes Jeremy can buy. $35 - 32 = 3$. Jeremy can buy 3 more tapes to fill the last travel case. Answer choice **F** is correct.

15. **C** The rectangle is divided into 10 equal sections, and 6 sections of the rectangle are shaded. That means that the fraction of shaded sections is 6 out of 10, or $\frac{6}{10}$. Answer choice **C** is correct.

16. **G** This question asks you to find the number pattern and decide what number will be next. The numbers are decreasing, so subtract 36 from 42 to find the difference. $42 - 36 = 6$. Next, find the difference between 36 and 30. $36 - 30 = 6$. Finally, find the difference between 30 and 24, because you want to make sure that 6 is always the difference. $30 - 24 = 6$. Good, 6 is the difference each time. To find the next number in the pattern, subtract 6 from 24. $24 - 6 = 18$. Answer choice **G** is correct.

17. **D** This is a good question to get rid of wrong answer choices. You can get rid of some answer choices just by looking at the first shape in the choices. The first shape in Cecil's line is a trapezoid. A trapezoid has exactly 1 pair of parallel lines. Look at the answer choices. Answer choice **A** has a pentagon as the first shape. That's wrong. Get rid of it. The first shape in answer choice **B** is a triangle. That's also wrong, so get rid of it too. The first shape in answer choices **C** and **D** is a trapezoid. Move on to the next shape in the line. The next one in Cecil's line is a square. In answer choice **C**, the next shape is a pentagon. That's not right. Get rid of it. That leaves answer choice **D.** Check the rest of the shapes to make sure they match.

18. **F** A prime number is a whole number that is greater than 1 and can only be divided by itself and 1. Answer choice **F** is $21 \div 7$, which equals 3. This is a prime number, because 3 can only be divided by itself and 1. It looks like the correct answer. Check the others to be sure. Answer choice **G:** $13 - 3 = 10$, which is not a prime number. Answer choice **H:** $17 + 5 = 22$, which is not a prime number. Finally, answer choice **J:** $5 \times 7 = 35$, which also is not a prime number. The correct answer is answer choice **F.**

19. **D** The number line goes from 8 to 10, with marks in between. Each mark stands for a tenth of a centimeter. The marks identify 8.1, 8.2, 8.3, 8.4, and so on. To find the point for 9.8, you can count 2 tenths to the left of 10 or count up from 8. This is arrow **D.**

20. **F** To find the number that fits in the box, solve the right side of the inequality first. $12 \div 3 = 4$. The left side of the inequality has to be *less* than 4. Substitute the answer choices into the box to find the correct answer. Answer choice **F:** $6 \div 2 = 3$, which is less than 4. This looks correct. Check the rest to make sure. Answer choice **G:** $8 \div 2 = 4$. This is not less than 4, so it's wrong. The remaining answers are larger, so they won't be less than 4 either!

21. **A** For this question, you will need to find the answer choice that is evenly divisible by 8. Try answer choice **A:** $32 \div 8 = 4$. It looks correct because 32 is evenly divisible by 8! But check the other choices, just to be sure. Answer choices **B** ($36 \div 8 = 4$ r4), **C** ($38 \div 8 = 4$ r6), and **D** ($42 \div 8 = 5$ r2) all have remainders. They can't be right. Answer choice **A** is correct.

22. **H** There are 6 numbers on the number cube. Three numbers are less than 4: 3, 2, and 1. Therefore, there is a 3 in 6 chance, or $\frac{3}{6}$, of throwing a number less than 4. **H** is the correct answer choice.

23. **D** This is a reasoning question. Sometimes you can find the answer by substituting actual numbers into the problem. Take the problem step by step. Zora shot more baskets than Nell. Give Zora 8 baskets and Nell 6 baskets, for example. Next, Nell shot more baskets than Ramiro. Give Ramiro 4 baskets. Now you can order them: $4 < 6 < 8$, or Ramiro $<$ Nell $<$ Zora. The opposite is also true: $8 > 6 > 4$, or Zora $>$ Nell $>$ Ramiro. See which answer choice has that information. Answer choice **A** is wrong, because Ramiro shot the fewest baskets. Get rid of answer choices **B** and **C** too. That leaves answer choice **D,** which is correct because Ramiro shot fewer baskets than Zora.

24. **G** Think about what happens when you multiply two numbers. When you multiply 9×6, you are finding 6 groups of 9. This is the same as adding 9 to itself 6 times, or $9 + 9 + 9 + 9 + 9 + 9$. Answer choice **G** is correct. Reminder: 9×6 is also the same as adding 9 groups of 6, or $6 + 6 + 6 + 6 + 6 + 6 + 6 + 6 + 6$.

25. **A** This question asks you to find the drawing that is made up of exactly 3 squares and 1 parallelogram. Look at each answer choice. Answer choice **A** seems to have 3 squares and 1 parallelogram. It could be right. But try the others, just to be sure. Answer choice **B** has 3 squares, but it also has 1 trapezoid, so it's incorrect. Get rid of it. Answer choice **C** has 2 parallelograms and 1 square. Get rid of it. The shapes in answer choice **D** are 2 squares and 2 trapezoids—not what you're looking for. The shapes you want are in answer choice **A**.

26. **J** There are 5 blocks in a bag, so there are 5 possible outcomes. The block colors are 1 white, 2 gray, and 2 black. Chet would have a 2 out of 5 chance of choosing a gray block. Answer choice **J** is correct.

27. **C** Use your pattern blocks when reading each answer choice. Answer choice **A** says that pattern block C is half the size of D. That's not right. Get rid of answer choice **A.** Try fitting pattern block C into block E. It can fit 3 times, so pattern block C is $\frac{1}{3}$ the size of E. This choice is wrong too. Fit pattern block C into block B. It fits twice, so block C is $\frac{1}{2}$ the size of block B. Answer choice **C** is correct. Pattern block A is more than 4 times larger than block C, so answer choice **D** must be wrong.

28. **H** You need to find out how many 6-inch cubes Kenyon can stack and still be taller than the tower. Kenyon is 57 inches tall. First, divide 57 by 6 to find how many cubes it will take to reach Kenyon's height. $57 \div 6 = 9$ r3. To reach Kenyon's exact height, he would need 9 cubes plus a part of another one. To remain shorter than Kenyon, the tower can be at most 9 cubes tall. Answer choice **H** is correct.

29. **B** To find the answer to this question, count the number of spaces on the grid that the vine grows. When it enters the grid, it grows 2 spaces up, then moves 2 spaces to the right, then 2 spaces up, then 2 spaces to the right. Now you've got the pattern: 2 up, 2 to the right. Continue that pattern to find where it will eventually reach. If it continues its current pattern, the vine will reach point B, which is answer choice **B**.

30. **H** There are 11 index cards, so there are 11 possible outcomes to this question. You know that there are 5 even numbers from 1 to 11: 2, 4, 6, 8, and 10. That gives Tabitha a 5 out of 11 chance of choosing an even-numbered card. Answer choice **H** is correct.

Session 2

31. This is a two-part question. Remember to show your work, especially when the question asks for it. If you do not show your work, you will lose credit, even if you have a correct answer.

Part A This part of the question asks you to measure sides of the stencil. The correct measurements appear below.

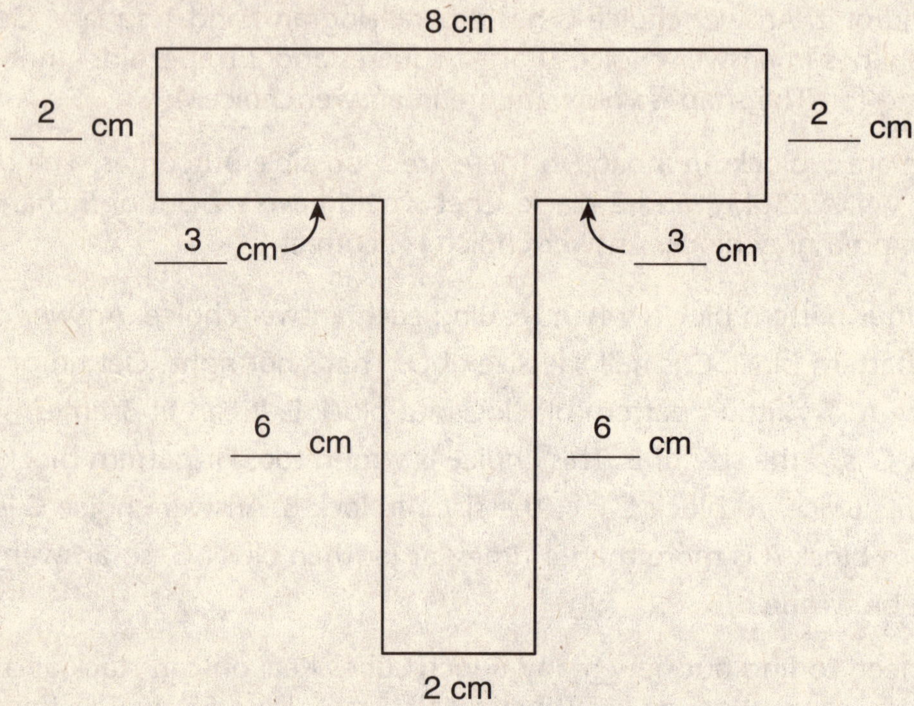

Part B To find the perimeter of a figure, you add all of the measurements. Under "Show your work," you would put

$$8 + 2 + 3 + 6 + 2 + 6 + 3 + 2 = 32$$

On the line next to "Answer," you would write 32.

You would receive full credit for the problem if you correctly filled in the measurements around the figure, showed your work to find the perimeter, *and* put the correct answer on the blank line. If you did not show your work and everything else was correct, you would lose a point. You would also receive some credit if you showed your work but got the incorrect answer.

If you want more help with finding perimeters, review Mile 24.

32. The equation uses the associative property, which means that the grouping of three or more factors can be changed without changing the product. In other words, no matter which two factors are multiplied first in the parentheses, the final product will be the same, so both sides are equal.

A good answer to put on the lines would be: Both sides of the number sentence are equal because of the associative property. Changing the grouping of three numbers will not change the answer.

You may still receive full credit if you explain the rule but don't identify it as the associative property.

33. Read this question very carefully so that you can answer it correctly.

For breakfast, Samuel ate $\frac{1}{3}$ of 12 grapes. $\frac{1}{3}$ of $12 = 4$. He ate 4 grapes for breakfast. For lunch he ate $\frac{1}{2}$ of the remaining grapes. Find how many grapes are left. $12 - 4 = 8$. Then find how many he ate. $\frac{1}{2}$ of $8 = 4$. He ate 4 grapes for lunch. To find out how many grapes he ate for breakfast and lunch, add the two answers. $4 + 4 = 8$. Samuel ate a total of 8 grapes.

34. This is a two-part question. Remember to show your work, especially when the question asks for it. If you do not show your work, you will lose credit, even if you have a correct answer.

Part A You've been asked to find the bottle that contains an amount closest to $740\frac{3}{4}$ ml. $\frac{3}{4} = 0.75$. Look at the contents of each bottle. Purity contains 740.60 ml, which is 0.15 ml too low. Springlike contains 740.70 ml, which is 0.05 ml too low. So far, this is the closest amount. The next bottle, Waterfresh, contains 740.85 ml, which is 0.10 ml too much. Springlike is still the closest. The last bottle, Good Tasting, contains less than Purity, so it is not correct. 740.70 ml is the closest to $740\frac{3}{4}$ ml. Springlike is the correct answer.

Part B Put the numbers in order from least to greatest.

$740.45 < 740.60 < 740.70 < 740.85$

Good Tasting < Purity < Springlike < Waterfresh

Purity has more water than Good Tasting but less than Springlike. The correct answer is Purity.

35. Use the grid to graph the information from the table. To get full credit for this question, you need to title the graph, label the axes, graph all the data, and use an appropriate scale. If you left out any of those items from your bar graph, you would lose credit. Below is a bar graph that would receive full credit.

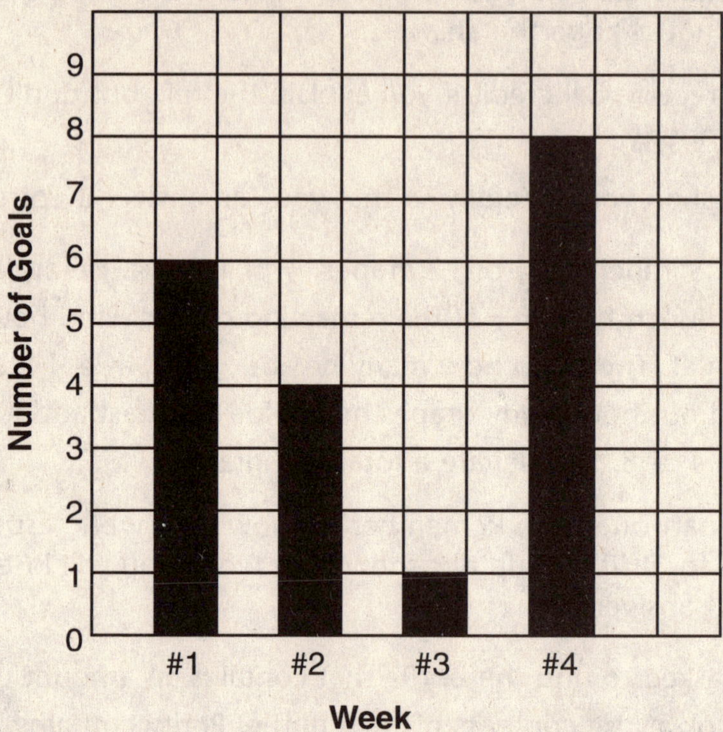

If you want more help with bar graphs, review Mile 29.

36. You can find the number of drawers in the chest by multiplying the number of drawers across by the number of drawers down. There are 5 drawers across and 8 drawers down. The equation is $5 \times 8 = 40$.

To write the division equation, take the product and divide it by one of the factors. You could write $40 \div 8 = 5$ or $40 \div 5 = 8$. Both answers are correct.

37. This is a two-part question. Remember to show your work, especially when the question asks for it. If you do not show your work, you will lose credit, even if you have a correct answer.

Part A To find out how much LaTasha would need to save each week, divide $28 by 4. 28 ÷ 4 = 7. LaTasha would need to save $7 per week.

If LaTasha wants to buy the present in 2 weeks, divide $28 by 2. $28 ÷ 2 = $14. She would need to save $14 per week.

Part B This part of the question asks if LaTasha could save the same amount of money for 3 weeks and buy the present. Divide $28 by 3 to find approximately how much LaTasha would have to save each week. 28 ÷ 3 = 9 r1.

If LaTasha saved $9 per week for 3 weeks, she would only save $27, which is not enough to buy the present. But if LaTasha saved $10 per week, she would have $30, more than enough to buy the present. A good answer for this question would be: LaTasha can save $10 per week for 3 weeks and have enough money to buy the present. She would have $2 left over.

38. To find the number of tiles it would take to cover Ms. Webster's kitchen floor, put your counters down over the floor space. You will use 9 counters. Ms. Webster needs 9 tiles for her floor.

Each counter equals 5 square units. To find the area of the floor, multiply the number of tiles by the number of square units of each counter. 9 × 5 = 45. The area of the floor is 45 square units.

39. This is a two-part question. Remember to show your work, especially when the question asks for it. If you do not show your work, you will lose credit, even if you have a correct answer.

Part A You are given the number of students in a school and the number of students in one grade. To find how many students are in all of the other grades, use subtraction. 891 − 107 = 784. There are 784 students not in Sam's grade in school.

Part B To find how many students are in each of the 7 remaining grades, use division. Take the number of students not in Sam's grade and divide it by the 7 other grades. 784 ÷ 7 = 112. There are 112 students in each of the 7 other grades.

Remember to show your work for this problem. The "Show your work" portion of the question could look like this:

```
  81
  891          112
 -107       7 )784
  784         -7
               08
               -7
               14
              -14
                0
```

SESSION 3

40. This is a two-part question. Remember to show your work, especially when the question asks for it. If you do not show your work, you will lose credit, even if you have a correct answer.

Part A First, find out how many roses are not red. Subtract the number of red roses from the total number Dermott gave his mother. $12 - 4 = 8$. There are 8 yellow or white roses. Next, find out how many are yellow. Of the 8 roses, there are 3 times as many yellow ones as white ones.

There are two ways that you can find the answer. You could take 8 counters and divide them into two piles. When one pile has 3 times more than the other pile, you'll have the correct answer. Or you could make a table with two columns showing the possible number of flowers.

yellow	white
1	7
2	6
3	5
4	4
5	3
6	2
7	1

Look at the columns to find which number of yellow roses is 3 times more than the number of white roses. That will be when there are 6 yellow roses and 2 white roses.

Part B Your explanation can use either of the methods above. If you have thought of another way to find the answer, write it down. You only have to explain how you found the answer.

Remember, giving the correct answer will earn you some of the points. Writing how you got that answer will earn you the rest of the points.

41. Look at each spinner to see what the probability would be for each spinner of the arrow landing on a vowel.

The vowels in the first spinner form about $\frac{1}{3}$ of the entire spinner. The consonant is $\frac{2}{3}$ of the spinner. Beth Ann would have a 1 out of 3 chance of the arrow landing on a vowel.

The vowels in the second spinner are $\frac{1}{2}$ of the spinner. Beth Ann would have a 1 out of 2 chance of the arrow landing on a vowel. This is a higher probability than the first spinner. Because the question is asking for the *least* probability, the first spinner is the better choice so far.

The vowels in the third spinner are $\frac{3}{4}$ of the spinner. Beth Ann would have a 3 out of 4 chance of the arrow landing on a vowel. This is the highest probability to land on a vowel of all of the spinners, so it is not correct.

The first spinner will give Beth Ann the least probability of landing on a vowel.

Your explanation could say: The first spinner has a 1 out of 3 probability of landing on a vowel, which is the least probability of the three spinners. Spinner 2 has a 1 out of 2 probability and Spinner 3 has a 3 out of 4 probability of landing on a vowel.

If you want to review probability, go to Mile 30.

42. This is a two-part question. Remember to show your work, especially when the question asks for it. If you do not show your work, you will lose credit, even if you have a correct answer.

Part A Each set of circles forms a square shape with the same number of circles on each side. The first set contains 1 circle on each side, or 1 circle in total. The second set contains 2 circles on each side, 4 circles in all. The third set contains 3 circles on each side, 9 circles in all. The next set in the pattern will have 4 circles on each side of the square for a total of 16 circles. When you draw the pattern, it will look like the figure below. Write the number 16 underneath it, or you will not get full credit.

16

Part B Each set of circles in this pattern forms a triangle with the same number of circles on each side. The first set has 1 circle. The second set has 3 circles, 2 on each side. The fourth set has 10 circles, 4 on each side. The third set will have 3 circles on each side for a total of 6 circles.

6

43. This is a two-part question. Remember to show your work, especially when the question asks for it. If you do not show your work, you will lose credit, even if you have a correct answer. Don't forget to use your counters!

Part A Use your counters to cover the entire letter. It will take 5 counters to cover the letter. If you use 3 counters, you will be covering 3 out of 5 parts, or $\frac{3}{5}$, of the letter. The correct answer is $\frac{3}{5}$.

Part B You discovered in Part A that it takes 5 counters to cover the letter. If you use 5 counters, then the letter is completely covered. So, the fraction of the shape that is *not* covered is $\frac{0}{5}$.

44. This is a two-part question. Remember to show your work, especially when the question asks for it. If you do not show your work, you will lose credit, even if you have a correct answer.

Part A To double the length of the square on the grid, count how long it is. It has a length of 3. You want the 3 to be doubled to a length of 6. Move the two lower points down 3 units. Connect the two new coordinates. The type of geometric shape is a rectangle.

Part B The ordered pairs of the new coordinates are (1, 1) and (4, 1).

If you would like to work more with grids and ordered pairs, go to Mile 27.

45. This is a two-part question. Remember to show your work, especially when the question asks for it. If you do not show your work, you will lose credit, even if you have a correct answer.

Part A Pattern block A is a hexagon. To make it using pattern block E, trace around it twice.

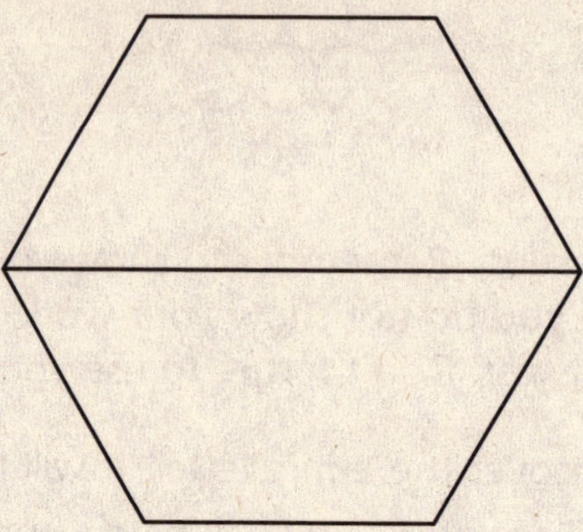

Part B See the diagram below. To get the correct pattern, you would have used pattern block E 8 times.

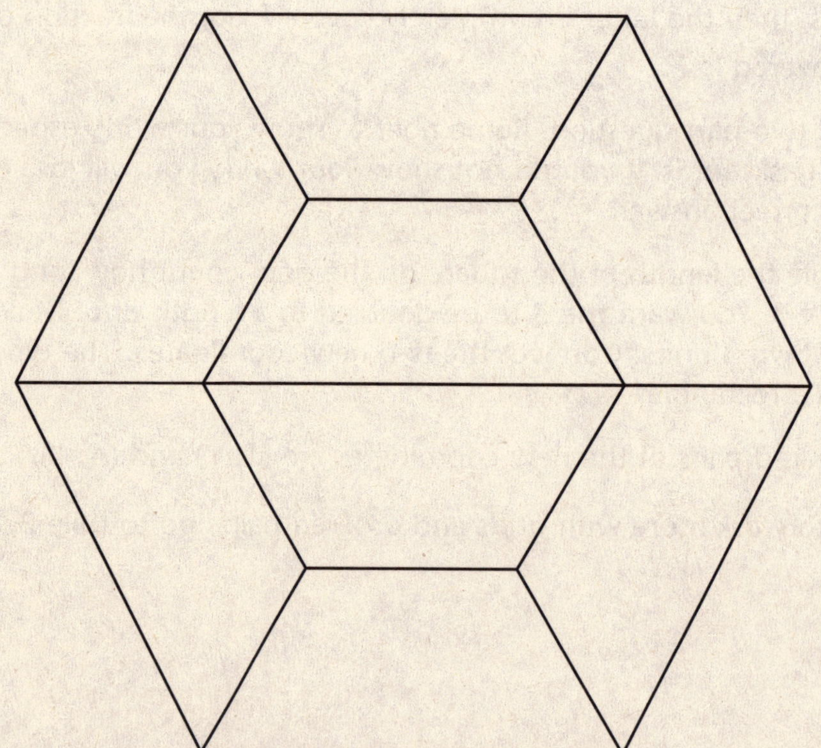

46. You need to find out how many pieces of sports equipment will be painted. Take it one step at a time.

There are 3 walls that will each have 22 paintings of athletes. Find the total number of athletes that will be painted by multiplying the number of walls by the number of athletes. $3 \times 22 = 66$.

Next, find the number of pieces of sports equipment that will be painted. Multiply the number of athletes, 66, by the number of pieces of sports equipment they will hold, 2. $66 \times 2 = 132$. There will be 132 pieces of sports equipment.

Remember: You must show your work for this problem or you will not receive full credit. You only have to put the two multiplication equations under "Show your work."

47. This is a two-part question. Remember to show your work, especially when the question asks for it. If you do not show your work, you will lose credit, even if you have a correct answer.

Part A The sign says that balloons cost \$0.40 each. To find how much 10 balloons will cost when purchased separately, multiply. $10 \times \$0.40 = \$4.00$. The correct answer is \$4.00.

Part B Joacquin has \$4.40. To find how many balloons he can buy separately, divide \$4.40 by \$0.40. $\$4.40 \div \$0.40 = 11$. Joacquin can buy 11 balloons separately. The question asks what is the *greatest* number of balloons he can buy. So see how many balloons he can buy if he buys a group. The group of balloons is 10 for \$3.50. If Joacquin buys one group, he will receive change. $\$4.40 - 3.50 = \$0.90$. With \$0.90 he can buy 2 more balloons separately for \$0.40 each. That will give him 12 balloons. The *greatest* number of balloons that Joacquin can buy is 12.

48. Use the diagram to help you solve this problem. You know that the sum of all four sides of the driveway is 52 feet and the width of the driveway is 12 feet. Label both widths 12 feet. The widths sum to $12 + 12 = 24$ feet. Subtract the widths from the perimeter. $52 - 24 = 28$ feet. Now that is the sum of the *two* lengths. To find one length, divide by 2. $28 \div 2 = 14$ feet. The length of the driveway is 14 feet.

Answers and Explanations for Practice Test 2

Session 1

Sample A **C** Don't forget to carry for this problem. The correct answer is 465. If you don't carry, you might end up with an incorrect answer choice, like choice **A.** Even if a problem on a test looks easy, take your time. You will receive a higher score if you do.

Sample B **J** Remember, page numbers increase. The page that would come next after 639 is 640, or answer choice **J**. Don't be fooled by the incorrect answers. Take your time and think about what the question is asking you.

Sample C **B** Take pattern block C and compare it to the drawing. You will find that it takes 6 of pattern block C to cover the entire larger shape, which means that it will take 3 to cover $\frac{1}{2}$ of the shape.

1. **B** You will need to borrow for this subtraction problem. Be careful when you borrow. Make sure that you change the number and cross out neatly. If you borrowed correctly, you would get 372, or answer choice **B**.

2. **H** Remember to carry for this multiplication problem. The correct answer is 318, or answer choice **H**. If you didn't carry the 1, you may have gotten 308, which is incorrect. Be careful and take your time.

3. **D** Get rid of the wrong answer choices. When you multiply 6×4, you are finding 6 groups of 4 things: $4 + 4 + 4 + 4 + 4 + 4$. Or you are finding 4 groups of 6 things: $6 + 6 + 6 + 6$. Also, $6 \times 4 = 4 \times 6$. The only answer choice that is *not* equal to 6×4 is answer choice **D.**

4. **F** To find the answer choice that is *not* true about 591,004, get rid of the answer choices that *are* true. Answer choice **F** isn't true because the 4 is in the ones place. It looks like you've found your answer. Check the others to be sure. The 5 is odd and in the hundred thousands place, so choice **G** is true. The 9 is in the ten thousands place. Answer choice **H** is true. The 1 has a value of 1,000. Get rid of answer choice **J.** Answer choice **F** is correct.

5. **C** You may not know the exact length of a telephone pole. That's okay! You can get rid of answer choices that seem wrong. Answer choice **A** is 10 centimeters. That's way too small. Get rid of that answer. Ten decimeters is about the height of a table. Get rid of answer choice **B.** Ten meters could be possible, but look at the last answer choice. Ten kilometers is very, very long—much too long for a telephone pole. The correct answer choice is **C,** 10 meters.

6. **G** To find how many pennies there are altogether, multiply 9×50. The correct answer is 450, answer choice **G.**

7. **A** Try drawing to help you compare fractions. First, draw a circle or a rectangle and shade half of it. Using the same shape, make sketches for the answer choices and find the one that is less than $\frac{1}{2}$. Answer choice **A,** $\frac{2}{7}$, is less than $\frac{1}{2}$. You should see that the other answer choices are all larger than $\frac{1}{2}$.

8. **J** Get rid of some of the answer choices using the first shape on the left. Ms. Matthews has a rhombus first. See which answer choices do too. Answer choices **F** and **J** have a rhombus first. Get rid of **G** and **H.** Look at the last shape in the line. It's a parallelogram! Because **F** lists a triangle last, the correct answer must be answer choice **J.**

9. **A** To solve this problem, do it in steps. First, add together the numbers on the right side of the inequality. $3 + 8 + 11 = 22$. Now, add together the numbers on the other side of the inequality. $12 + 7 = 19$. Find the difference between the two sides. $22 - 19 = 3$. The left side of the inequality must be less than the right side, so you want a number that is less than 3. Answer choice **A** is 2. That is the correct answer.

10. **H** Newton earned 85 points out of 100, or $\frac{85}{100}$. That is the same as 85%, or answer choice **H.** If you want to review percentages, go to Mile 14.

11. **B** This question wants you to find how much of the cup is empty. The cup is divided into sixths. Two of the sixths are not shaded, which means that they are empty. The correct answer is $\frac{2}{6}$, or choice **B.**

12. **J** Each package of chalk has 12 pieces. Because Mr. Williams only bought complete packages, the total pieces of chalk that he bought must be a multiple of 12. Divide each answer by 12. The answer choice without a remainder must be correct. Answer choice **F:** $64 \div 12 = 5$ r4. Answer choice **G:** $68 \div 12 = 5$ r8. Answer choice **H:** $70 \div 12 = 5$ r10. Answer choice **J:** $72 \div 12 = 6$. The correct answer is 72, choice **J.**

13. **C** You'll need to find the hours and minutes of each answer choice to find the correct answer. Answer choice **A** is 10:15 P.M. to 8:00 A.M. It's 45 minutes from 10:15 to 11:00. From 11:00 to 8:00 is 9 hours. This answer choice is 9 hours and 45 minutes, which is wrong. Get rid of it. Choice **B** is 10:45 P.M. to 8:00 A.M., which is 30 minutes less than answer choice **A** and is not right. Get rid of it. Answer choice **C** is 11:15 P.M. to 8:00 A.M. It's 45 minutes from 11:15 to 12:00. From 12:00 to 8:00 is 8 hours. This answer choice is 8 hours and 45 minutes and is the correct answer. If you need more help working with time, review Mile 21.

14. **H** Pat's bag has 10 marbles, and 3 are gray. That would give Pat a 3 out of 10 probability of choosing a gray marble. That's incorrect. Get rid of it. Mel has 9 marbles in his bag, and 4 are gray. That gives Mel a 4 out 9 probability of choosing a gray marble. It's also incorrect. Rafi's bag has 9 marbles, and 3 are gray. Rafi has a 3 out of 9 probability of choosing a gray marble, which is the same as 1 out of 3. This looks correct, but check the last one. Carol's has 9 marbles, and 2 are gray. Carol has a 2 out of 9 probability of choosing a gray marble, which is not correct. Answer choice **H** is correct.

15. **B** The number pattern increases by a regular amount. What is added to each number to get the next number? 7! To find the number that starts the pattern, subtract 7 from 42. $42 - 7 = 35$. Answer choice **B** is correct.

16. **G** Compare your pattern blocks to answer this question correctly. For answer choice **F**, pattern block C would fit into pattern block A 6 times, so this answer choice is wrong. Pattern block E fits into pattern block A 2 times. Answer choice **G** looks right, but read the other answer choices to make sure. Pattern block A is the largest pattern block, so it cannot be $\frac{1}{2}$ the size of any other pattern block. Answer choice **G** is correct.

17. **A** You can substitute numbers for each bag of apples to help find the correct answer. Bag R weighed less than T. Make R weigh 3 pounds and T weigh 5 pounds. Bag S weighs less than R, so make it weigh 2 pounds. Now you have $2 < 3 < 5$ and $S < R < T$. Look decide which answer choice is correct. T does weigh more than S, so answer choice **A** is the correct answer.

18. **F** The most important part of this question is to know the definition of parallel lines. Parallel lines are lines that will never intersect. Now look at the figure. Does it have 2 lines that will never intersect? No. All of its lines will intersect, so it has no parallel lines. The correct answer is choice **F**.

19. **D** The question tells you that an eagle has a greater average wingspan than a pelican but a smaller average wingspan than a condor. The number line shows the pelican wingspan as 2.1 meters and the condor wingspan as 3.0 meters. The eagle's wingspan should be somewhere in between. The only answer choice between 2.1 and 3 is **D**, or 2.5 meters.

20. **G** The design described has 2 triangles and 1 trapezoid. Look at the figures and get rid of wrong answer choices. Answer choice **F** isn't correct because there's a square in the figure. Get rid of it. Answer choices **H** and **J** are made up of 3 triangles. Only answer choice **G** has 2 triangles and 1 trapezoid. That's the correct answer.

21. **A** Students leave and no new students come, so this is a subtraction problem. Take the total number of students and subtract the ones that move away. $595 - 19$ is the correct expression. If you need more help with word problems, review Mile 11.

22. **G** First, count the total number of sections on the spinner. There are 8. This is the total number of possible outcomes. There are 2 vowels. The probability that Tia's spinner will land on a vowel is 2 out of 8, or $\frac{2}{8}$. Answer choice **G** is correct.

23. **D** Divide the total number of cans by the number that will fit into each box. $35 \div 8 = 4$ r3. Jasmine had 4 complete boxes and 3 cans to put into the last box. The correct answer is 3, or answer choice **D**.

24. **H** This is an estimation problem. You don't need to get the exact answer. Round 5.9 centimeters to 6 centimeters. Now multiply by 10 and you'll get 60, or answer choice **H**. If you would like to practice more estimation problems, go to Mile 18.

25. **B** Find the bar for All Flavors on the graph. Then look to the left to read the price of a cone. It's $1.00, or answer choice **B**.

26. **H** One way to answer this question is to substitute actual numbers and do the math. Use 2 for the even number and 3 for the odd number. Answer choice **F** adds the two numbers. 2 + 3 = 5. That's an odd sum, so it's not the correct answer. Answer choice **G** would be 2 + 2 + 3 = 7. That's also an odd sum and not the right answer. Answer choice **H** is 3 + 3 = 6. That is an even sum and seems to be the correct answer. Look at answer choice **J**. 3 + 3 + 3 = 9. No, that's not right either. Answer choice **H** must be the correct answer.

27. **A** Look at the pattern on the grid. When Stan and Jerry enter the grid, they move 3 spaces down, and then they move 2 spaces to the left. They go 3 spaces down again and 2 spaces to the left. This is the pattern that you need to continue along the grid. If you follow 3 spaces down and 2 spaces to the left, you will meet point A, which is answer choice **A**.

28. **J** Find the answer one step at a time. There are 9 gerbils and 4 fewer puppies than gerbils. To find the number of puppies, subtract. 9 − 4 = 5 puppies. There are 6 more kittens than puppies. To find the number of kittens, add. 5 + 6 = 11 kittens. There are 11 kittens, answer choice **J**.

29. **C** Find the probability by first calculating the total number of possible outcomes. Add all of the T-shirts in the drawer. 2 + 3 + 2 = 7. There are 7 possible outcomes. There are 3 red T-shirts, so there is a 3 out of 7 or $\frac{3}{7}$ chance that Jonathan will pick a red T-shirt from the drawer. Answer choice **C** is correct.

30. **H** You can see from the diagram that there are 56 inches of ribbon. Divide 56 by 6 to find how many equal pieces Ivana can get. 56 ÷ 6 = 9 r2. She can get 9 pieces and have a little bit left over, but she doesn't have enough ribbon to make 10 equal pieces. The correct answer is answer choice **H**, 9.

Session 2

31. To find the answer to the first question, look at the two cards that are whole. The first card shows a relationship between 72, 8, and 9. $72 \div 8 = 9$, so the missing symbol might be a division sign. Try the other card to be sure. $44 \div 11 = 4$. The symbol must stand for division.

    One card is missing a number, but you can see the same symbol. Based on the other cards, the symbol must be division. The missing number is 42, because $42 \div 6 = 7$.

32. The question asks you to show your work. If you don't show your work, you may not get full credit, even if you have the correct answer. Remember, you can get some credit for showing work on the problem, even if you don't get the correct answer.

    You may use your counters to solve this problem. Separate the counters into groups that represent the number of houses Mr. Pauling sold each month. The first group, January, should have 2 counters. He sold twice that many in February, so the next group should have 4 counters. In March, Mr. Pauling sold $1\frac{1}{2}$ times the number in February. The next group of counters should have 4 to represent the same amount as February *and* 2 more to represent half. There should be 6 counters in this group. Count up all the counters in all three groups to find the total number of houses Mr. Pauling sold. It's 12.

33. This is a two-part question. Remember to show your work, especially when the question asks for it. If you do not show your work, you will lose credit, even if you have a correct answer.

Part A Randy added 8 ounces of water to the fish tank each week. To find how much he has added at the end of 9 weeks, multiply. $8 \times 9 = 72$. He added 72 ounces.

To find the amount he added after 14 weeks, multiply. $14 \times 8 = 112$. He added 112 ounces.

Part B First, find out how much water would be lost after 18 weeks. $8 \times 18 = 144$ ounces. A gallon of water is equal to 128 ounces, which is less than the 144 ounces.

A good answer would be: If Randy did not add water to the tank for 18 weeks, the tank would have lost 144 ounces. That is more than 1 gallon of water, 128 ounces.

34. Without doing any calculations, you can know that both sides of the number sentence are equal because of the commutative property.

A good answer would be: The commutative property says that you can change the order of two or more addends without changing the sum. Therefore, the number sentence is true.

You may still receive full credit if you explain the rule but do not call it the commutative property.

35. This is a two-part question. Remember to show your work, especially when the question asks for it. If you do not show your work, you will lose credit, even if you have a correct answer.

Part A The measurement of each side of the box is 9 centimeters. Write 9 on each of the blank lines.

Part B The perimeter of a figure is the sum of all of its sides. The perimeter of this triangular box would be $9 + 9 + 9 = 27$ centimeters.

To get full credit for this part of the question, you would have had to show the equation under "Show your work." You would also have had to write the number 27 on the blank line.

36. Put your counters on the drawing of the shed. It should have taken 6 counters to cover the wall of the shed. If 1 cup of paint can cover the area of one counter, it would take 6 cups to paint the side of the shed.

Multiply the number of counters by the area. $16 \times 6 = 96$. The area is 96 square units.

37. This is a two-part question. Remember to show your work, especially when the question asks for it. If you do not show your work, you will lose credit, even if you have a correct answer.

Part A To find the box that weighs the nearest to $\frac{7}{10}$ ounce, rewrite $\frac{7}{10}$ in decimal form, 0.7. The two boxes that are closest in weight to 0.7 ounce are Box 1 and Box 3. The difference in weight between Box 1 and 0.7 ounce is $0.80 - 0.70 = 0.1$ ounce. The difference in weight between Box 3 and 0.7 ounce is $0.70 - 0.67 = 0.03$ ounce. Remember to use 0 as a placeholder when calculating with decimals. Since 0.03 is less than 0.1, Box 3 is the nearest to $\frac{7}{10}$ ounce.

Part B Box 3 weighs 0.67 ounce and Box 4 weighs 1.02 ounces. Box 1 weighs 0.80 ounce. $0.67 < 0.80 < 1.02$. Box 1 is the correct answer.

38. This is a two-part question. Remember to show your work, especially when the question asks for it. If you do not show your work, you will lose credit, even if you have a correct answer.

Part A The height of the shorter redwood tree is 172 feet less than the height of the tallest redwood. Subtract the difference from the height of the tallest redwood. $368 - 172 = 196$. The shorter redwood tree is 196 feet.

Part B This question needs to be done in two steps. Read the first sentence. A maple tree in Smith Park is 4 times shorter than the tallest redwood tree. To find the height of the maple tree, divide the height of the tallest redwood by 4. $368 \div 4 = 92$. The maple tree is 92 feet tall.

The maple tree in Smith Park is 92 feet. The question says that an average maple tree is 100 feet. Subtract to find the difference. $100 - 92 = 8$. The tree in Smith Park is 8 feet shorter than an average maple tree.

39. Use the grid to graph the information from the table. To receive full credit for the bar graph, you need to title the graph, label the axes, graph all the data, and use an appropriate scale. The complete graph would look as follows:

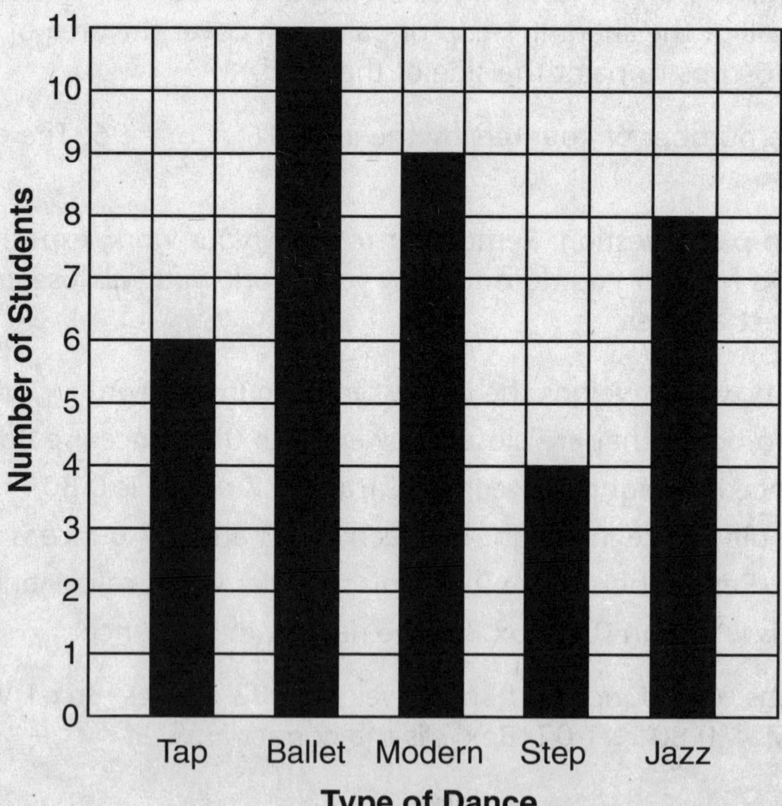

If you want to practice more with bar graphs, go to Mile 29.

Session 3

40. To answer this question, look at each square and find the probability of a coin landing on an even number.

    Square 1: The area of the square covered by the two even numbers is half of the square. The probability of a coin landing on an even number is $\frac{1}{2}$.

    Square 2: The largest section, 1, takes up half the square. The odd numbers 1 and 3 have a larger combined area than in Square 1, so there's a larger probability that the coin will land on an odd number. Square 1 is a better choice so far.

    Square 3: The even-numbered sections make up more than half of the area of this square. The probability of a coin landing on an even number is greater than $\frac{1}{2}$. This square has the greatest probability of having a coin land on an even number.

    Square 3 has the greatest probability because the even-numbered sections take up the greatest area.

41. This is a two-part question. Remember that it's important to show your work, especially when the question asks for it. If you do not show your work, you will lose some credit, even if you have a correct answer.

Part A Take pattern block C and place it over the other pattern blocks. See which one is twice its size. You will find that pattern block B is the correct answer.

Part B If pattern block C has an area of 1, then a rhombus with an area of 8 would be equal to 8 of pattern block C. Remember that a rhombus is a four-sided figure with opposite sides parallel and equal in length. You should have rearranged your shapes until you found a figure that met all the requirements. This is drawn below.

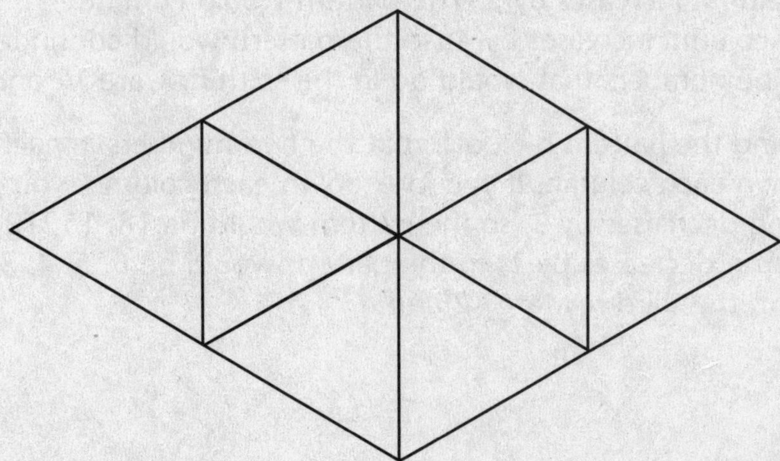

42. This is a two-part question. Remember to show your work, especially when the question asks for it. If you do not show your work, you will lose credit, even if you have a correct answer.

Part A Niles can only buy 4 sheets of poster paper separately. Multiply the price of 1 sheet of poster paper by 4 to find how much it will cost. $4 \times \$0.85 = \$3.40$. A value pack costs \$3.75. Subtract to find how much less Niles will spend. $\$3.75 - \$3.40 = \$0.35$.

Part B You can find the price of 8 sheets of poster paper by multiplying the price of 1 sheet by 8. $8 \times \$0.85 = \$6.80$. Buying the sheets separately will cost \$6.80. But there is a value pack of 5 sheets. Niles's sister could buy a value pack and 3 additional sheets for a total of 8 sheets. Add the cost of the value pack to the cost of 3 sheets. $\$3.75 + (3 \times \$0.85) = \$3.75 + \$2.55 = \$6.30$. This is less than buying the sheets separately. The least Niles's sister could spend is \$6.30.

43. This is a two-part question. Remember to show your work, especially when the question asks for it. If you do not show your work, you will lose credit, even if you have a correct answer.

Part A Carefully put your counters on $\frac{3}{4}$ of the rectangle. The length of the rectangle is 4 counters, and the width is 3 counters. When you cover $\frac{3}{4}$ of them, you will use 9 counters.

Part B Take away 3 of the counters to have 6 left on the rectangle. You will now be covering $\frac{1}{2}$ of the rectangle with the counters.

44. This is a two-part question. Remember to show your work, especially when the question asks for it. If you do not show your work, you will lose credit, even if you have a correct answer.

Part A You can find the pattern by looking at the two numbers in each row across or by going down each column. If you look down each column, you will see that the left column increases by 2. The pattern would continue 1, 3, 5, 7, 9, and 11. The right column increases by 4, so the pattern would continue 2, 6, 10, 14, 18, and 22. The numbers that would go in the sixth row are 11 and 22.

Part B You can find the pattern by looking at the two numbers in each row across or by going down each column. If you look down each column, you will see that the left column decreases by 3, so the pattern would be 18, 15, 12, 9, 6, and 3. The right column decreases by 1, so the pattern would be 6, 5, 4, 3, 2, and 1. The numbers in the third row are 12 and 4.

45. Ms. Kipfer makes 3 designs of sun catchers and sends 15 of each design to each store. That means she sends $3 \times 15 = 45$ sun catchers to each store. There are 4 different stores, so Ms. Kipfer sent a total of $45 \times 4 = 180$ sun catchers.

Don't forget to show your work. If you had written each equation in the open space, solved it, and then written the correct answer on the blank line, you would have received full credit.

46. This is a two-part question. Remember to show your work, especially when the question asks for it. If you do not show your work, you will lose credit, even if you have a correct answer.

Part A Look at the figure on the graph. It has four sides and one pair of parallel lines. It's a trapezoid.

Part B Remember that ordered pairs list the distance to right first, then the distance up. The lower left point is 1 unit to the right of zero and 2 units up. Its ordered pair is (1, 2). The lower right corner of the trapezoid is at (8, 2). The other two ordered pairs are (3, 6) and (6, 6).

If you want to review finding ordered pairs, go to Mile 27.

47. This is a two-part question. Remember to show your work, especially when the question asks for it. If you do not show your work, you will lose credit, even if you have a correct answer.

Part A To find one of the dimensions, add the measures of all the sides that go across. $4 + 4 + 4 = 12$ feet. The width of the garden is 12 feet. Now add the measures of the lines that go up and down. Be careful, though, because you only want to add them once. $9 + 3 = 12$. The length of the garden is also 12 feet.

Part B To find the perimeter of the garden, add the measurements of each side of the garden. The width, which is the bottom side, is 12 feet.

$12 + 3 + 4 + 9 + 4 + 9 + 4 + 3 = 48$ feet

The perimeter of the garden is 48 feet.

48. This is a two-part question. Remember to show your work, especially when the question asks for it. If you do not show your work, you will lose credit, even if you have a correct answer.

Part A Raphael had 14 markers. Subtract the 3 yellow markers and the 2 green markers to find the number of red or blue markers. $14 - 3 = 11$ and $11 - 2 = 9$. There were 9 red and blue markers. You could have made a list of the possible combinations of the colors.

red	blue
1	8
2	7
3	6
4	5
5	4
6	3
7	2
8	1

Remember, there were half as many red markers as blue markers. Which pair of numbers shows this relationship? Look at the list. When there are 3 red markers, there are 6 blue markers. Raphael had 3 red markers.

Part B Explain how you found your answer. You could describe the table above, or you could have used your counters. Just be sure to explain exactly how you did it.

One good answer would be: I took 9 counters and put them into two different piles. When I had 6 in one pile, I found that I had half of that, or 3, in the other pile.

Find the Right School

Best 351 Colleges
2004 Edition
0-375-76337-6 • $21.95

Complete Book of Colleges
2004 Edition
0-375-76330-9 • $24.95

Complete Book of
Distance Learning Schools
0-375-76204-3 • $21.00

America's Elite Colleges
The Smart Buyer's Guide to the Ivy League and Other Top Schools
0-375-76206-X • $15.95

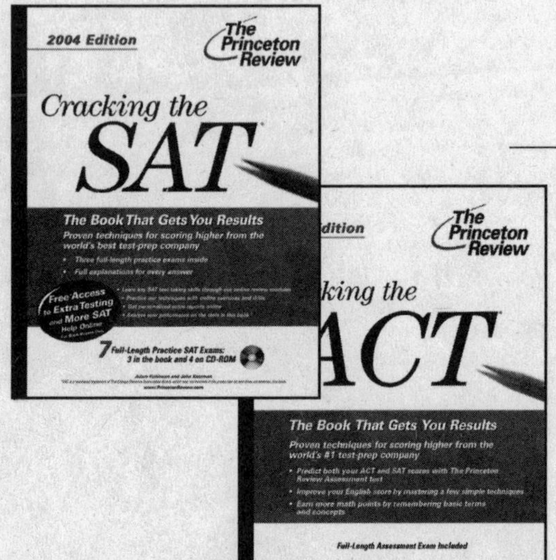

Get in

Cracking the SAT
2004 Edition
0-375-76331-7 • $19.00

Cracking the SAT
with Sample Tests on CD-ROM
2004 Edition
0-375-76330-9 • $30.95

Math Workout for the SAT
2nd Edition
0-375-76177-2 • $14.95

Verbal Workout for the SAT
2nd Edition
0-375-76176-4 • $14.95

Cracking the ACT
2003 Edition
0-375-76317-1 • $19.00

Cracking the ACT with
Sample Tests on CD-ROM
2003 Edition
0-375-76318-X • $29.95

Crash Course for the ACT
2nd Edition
The Last-Minute Guide to Scoring High
0-375-75364-3 • $9.95

Crash Course for the SAT
2nd Edition
The Last-Minute Guide to Scoring High
0-375-75361-9 • $9.95

Get Help Paying for it

Dollars & Sense for College Students
How Not to Run Out of Money by Midterms
0-375-75206-4 • $10.95

Paying for College Without Going Broke
2004 Edition
0-375-76350-3 • $20.00

The Scholarship Advisor
5th Edition
0-375-76210-8 • $26.00

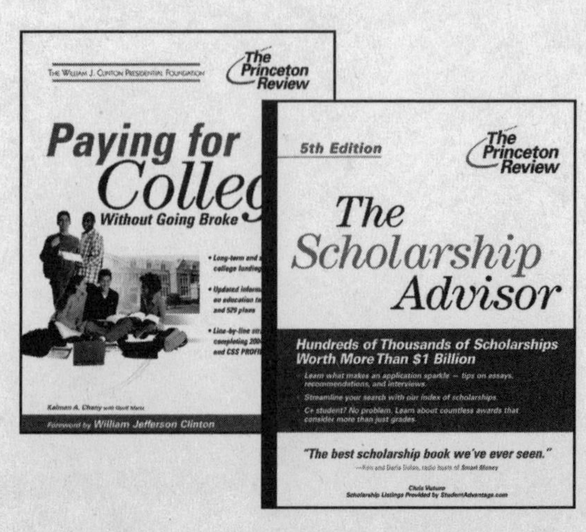